EndNote® 1-2-3 Easy!

Reference Management for the Professional

EndNote® 1-2-3 Easy!

Reference Management for the Professional

Abha Agrawal
Kings County Hospital Center
Brooklyn, NY

Abha Agrawal, MD
Director, Medical Informatics
Associate Medical Director
Kings County. Hospital Center
451 Clarkson Ave.
Brooklyn NY 11203
USA

EndNote is a registered trademark of The Thomson Corporation.

Library of Congress Cataloging-in-Publication Data: 2005923025

ISBN-13: 978-0387-24991-9 e-ISBN: 978-0-387-25491-3

Printed on acid-free paper.

Printed in the United States of America.

9 8 7 6 5 4 3

springer.com

For Vijaya

Preface

About the Book

Reference management is the process of storing, managing, retrieving, and citing references from various sources. Learning effective reference management is critically important for healthcare and biomedical professionals as it is a major component of the process of scientific manuscript writing. Effective reference management is also useful for other purposes such as maintaining a list of scientific references on a topic of interest, delivering presentations, preparing for journal clubs or case conferences, and teaching activities.

Scientific manuscript writing is an extremely common task. Tens of thousands of articles are added per month just to one biomedical database—MEDLINE. Even this is a tiny fraction of the actual volume of the writing activity as most of what is written does not get published. In addition, reference management is useful for writing reports, grants, and other non-journal articles as well.

A number of reference management software programs are available these days to make this challenging task simpler by harnessing the power of information technology. EndNote® is one of the most popular software among biomedical and healthcare professionals.

The purpose of this book is to educate readers to effectively and efficiently use EndNote. The book provides step-by-step instruction on using EndNote to create a personal digital library of scientific references and to create an accurately formatted bibliography in a manuscript.

The book will help novice users in rapidly learning to perform important reference management tasks using EndNote. It will enhance the skills of experienced users in performing advanced tasks such as using EndNote to retrieve references from Internet databases, electronic journals, or other e-resources. For nonusers of technology, I hope it will stimulate their interest in exploring the use of EndNote in their professional activities.

While overall EndNote is a user-friendly program, professionals often find that there is a steep learning curve that prevents them from trying to use it for their work. They often cannot get over the initial hassle and time investment of trying to learn the user interface and other intricacies of a new software program. Software manuals are generally cumbersome to read and are not specifically designed for the needs of healthcare and biomedical professionals.

The book is written keeping the requirements, skills, and the time constraints of healthcare and biomedical professionals in mind. All chapters in the book are task-oriented; you can simply use the "Table of Contents" or the "Index" to find the task you would like to perform without reading the entire chapter. I have used ample illustrations such as computer screen shots and flow diagrams to provide a visual display of how to perform various tasks.

The book derives from my extensive experience as a user of EndNote as well as from the practical insights gained by helping my colleagues over the years in the use of the software to get their manuscripts submitted in proper format and style. It stems from my understanding of the problems faced by novice users trying to learn a new program as well as expectations of more experienced users as they optimize the use of this application. I hope that sharing this experience with you will help you get over the initial steep learning curve of using EndNote.

Disclaimer: I do not endorse any company or its products. This book is solely based on my experience and objective analysis.

This Book is for You

- If you write manuscripts that include text references, figures, charts, or tables.
- If you submit articles to scientific journals and other publishers and would like to ensure the accurate formatting of the in-text citations and the bibliography.
- If you would like to minimize the time involved in renumbering and rearranging your bibliography during iterative improvements in the manuscript.
- If you search online databases such as PubMed, the Library of Congress, or the Web of Science and would like to store selected references in an electronic database.
- If you read scientific journals online and would like to automatically download citations to selected articles in an electronic library for future use.
- If you would like to generate a list of articles or other citations on topics of your interest for teaching or as reading lists.
- If you teach reference management or EndNote courses.

You Should Know

You do not need to have any prior knowledge of EndNote or any other reference management program. You do *not* need to be a skilled or experienced computer user. It would be helpful for you to have the knowledge of:

- Basic Windows operations such as opening, closing, and saving a file, minimizing and maximizing windows etc.
- Basic use of the keyboard and the mouse.

- Basic word processor (e.g. Microsoft Word) operations that you would need for writing your manuscript with or without EndNote.

This book is meant to be used with computers running Windows® Operating Systems only. Note that EndNote is available for Windows as well as MacIntosh® operating systems.

Conventions Used in This Book

Technical Tip: The tips will help you use a program more efficiently and will indicate to you features of the program that were not otherwise obvious.

Alert: When you see an alert, read it carefully—it will help you avoid possible problems in performing a task.

Typographic Conventions

Italics are used for menu command and other commands that will help you interact with the computers. For example, "*Edit > Preferences*" means you will click on "Edit" in the menu bar and then on "Preferences" in the submenu.

URLs

A word about URLs used in this book: I have taken extra caution to make sure that the URLs point to the correct web pages and the URLs are current as of the writing of this book. However, given the volatile nature of web publishing you may occasionally find that a URL is nonfunctional.

Acknowledgements

I gratefully acknowledge the help of Dr. James Reilly, Professor of Surgery at the State University of New York College of Medicine, Brooklyn, and Ms. Jan Glover, Education Services and Reference Librarian at the Cushing/Whitney Medical Library of Yale School of Medicine, for their critical review of the manuscript and constructive comments.

Contents

1. Introduction to Reference, Bibliography, and Citation 1
Basic Concepts 1
Reference 1
Reference List and Bibliography 2
Citation 3
Common Referencing Styles 3
Using Information Technology for Effective Reference Management 5

2. Reference Management Software Programs 7
What is a Reference Management Software Program? 7
Functions of Reference Management Software Programs 8
Storing and Managing References 8
Creating Bibliographies 9
Searching and Retrieving References from Online Databases 9
Working with a Handheld Computer 9
Various Reference Management Software Programs 10
Which Program Should You Choose? 13

3. Getting Started with EndNote 15
An Overview of Working with EndNote 15
Technical Requirements for Using EndNote 15
Hardware Requirements 16
Operating System (OS) Requirements 16
Word Processor Compatibility 17
Hand-Held Computer Requirements 17
Getting EndNote 18
Trial Version of EndNote 18
Installing EndNote 19
Installing EndNote for the First Time 19
Upgrading from an Earlier Version of EndNote 19
Installing EndNote onto a Network 21

Checking EndNote Installation 22
Checking EndNote 22
Checking Word Processor Support 23
Automatically Updating EndNote 33
Uninstalling EndNote 35
File Compatibility Issues 35
EndNote Library 35
EndNote Styles 35

4. EndNote Libraries 37
What is an EndNote Library? 37
Features of an EndNote Library 37
Working with an EndNote Library 38
Opening a Library 39
Understanding the EndNote Library Window (Figure 5) 40
Creating a New Library 41
Sorting the EndNote Library 42
Navigating the EndNote Library 43
Previewing a Reference 44
Copying Between Libraries 45
Setting Library Preferences 45
Setting Fonts in the Library 46
Setting a Default Library 47
Setting the Fields to be Displayed in the Library 48
Recovering a Damaged Library 49
Helpful Hints About a Recovered Library 50
Merging EndNote Libraries 51
Publishing an EndNote Library on the Web 52

5. Entering References into EndNote Library 57
Introduction 57
Reference Fields and Reference Types 58
Customizing Reference Types 58
Creating a New Reference 61
Choosing the Right Reference Type 62
Setting the Default Reference Type 62
Entering Reference Data Manually 63
Let EndNote Do the Formatting 63
Guidelines for Entering Data in Various Reference Fields 64
Term Lists 65
What is a Term List? 66
Two Basic Features of Term Lists 66
Journal Term Lists 66
Helpful Hints About Term Lists 68
Turning Off Term List Features 68

The 'Figure' and the 'Chart or Table' Type Reference 68
Creating a Figure Type Reference 69
Creating a Chart or Table Type Reference 69
Notes About Figure or Chart/Table Type References 70
Entering Special Characters in References 72
Spell-Checking 74
Downloading References from the Websites of Journals 75
Importing References from Other Reference Management Programs into EndNote 78

6. Managing References in an EndNote Library 79
Understanding the Reference Window 79
Working with References 80
Selecting References 80
Opening References 82
Saving References 82
Deleting References 82
Reverting References 83
Showing and Hiding References 83
Exporting References 84
Searching References 85
Launching EndNote Search 85
Understanding the Search Window 85
Performing Search 88
Duplicate References 88
Checking for Duplicate References 88
Customizing Settings for Find Duplicates Command 90
Deleting Duplicate References 90
Group Editing of References 91
The Change Text Command 92
The Change Fields Command 94
The Move Fields Command 94
Linking References to Files and Websites 96
Linking References to Files 96
To Link a File to a Reference 97
Linking References to Websites 98
OpenURL Link Command 98

7. Using EndNote with Internet Databases 101
Introduction 101
Various Methods of using EndNote with Internet Databases 102
Choosing the Right Connection Method 104
The Connection File Method 106
What is a Connection File? 106
What Connection Files do I Have? 106

Working with the Connection Manager 107
Setting "Favorite" Connection Files 107
Downloading Connection Files from the Internet 107
Using the Connection File Method 109
Working in the Retrieved Reference Window 112
The Import Filter Method 114
What is an Import Filter? 114
What Import Filters do I Have? 114
Using the Import Filter Method 116
The Direct Export Method 116
Using EndNote with PubMed® 117
Introduction 117
Using the Connection File Method for PubMed 118
Using the Import Filter Method for PubMed 120
Steps for Using the Import Filter Method for PubMed 120
Using EndNote with Ovid® 126
Introduction 126
Using the Direct Export Method for Ovid 126
Using EndNote with the Web of Science® 127
Introduction 127
Using the Direct Export Method for the Web of Science 129

8. Creating Bibliographies Using EndNote 131
An Overview of Steps in Using EndNote to Create Bibliographies . . . 131
Output Styles 133
The Style Manager 134
Working in the Style Manager 134
The Style Manager Window 135
Marking Styles as Favorites 135
Editing Styles 135
Examples of Editing Styles 137
Creating a Manuscript 139
Inserting References from an EndNote Library into a Manuscript . . . 140
Inserting References into Manuscript 140
Changing Existing Citations 142
Creating Bibliographies 144
Formatting Bibliography 144
Customizing the Bibliography 144
Finding and Editing Cited References in a Library 146
Creating a Bibliography from Multiple Documents 146
Including Notes in the List of References 148
Working with Figure and Table/Chart References in Manuscripts . . . 149
Introduction 149
Working with Figures and Tables/Charts 150
Inserting Figures and Tables/Charts in a Document 150
Customizing Figures and Tables/Charts in a Document 151

Sending Paper to Publisher/Sharing with Others 153
Field Codes 153
Traveling Library 155
Sharing Your Document with Others: Creating a Traveling Library 156
Sending Your Paper to a Publisher 157
Other Tasks 157
Citing References in Footnotes 157
Customizing Footnote Citations 159
Creating an Independent Bibliography 160
Creating a Subject Bibliography and Subject List 163
Setting CWYW Preferences 168
Miscellaneous Tasks 170

9. Citing References from Sources on the Internet 175
Introduction 175
What's Different about Citing Internet Sources? 176
General Principles for Citing Internet Sources 177
Guidelines for Citing Internet Sources 177
Authors 178
Title 178
URL 179
Edition 181
Dates 181
Page Information 182
In-Text Citations of Internet Sources 182
Examples of References from Internet Sources 182
Sources on the World Wide Web (WWW) 182
E-Mail Messages 184
Web Discussion Forum Posting 185
Listserv Message 185
Newsgroup Message 186
FTP Sites 186
Software Programs 186
Using EndNote to Manage References from Internet Sources 187
Customizing EndNote to Cite References from Internet Sources . . 188
Spelling and Definition of Commonly Used e-Terms 192

10. EndNote for PDA Computers 193
Introduction 193
Technical Requirements for Using EndNote on a PDA 194
Installing EndNote for Palm Application 196
Copying an EndNote Library from Desktop to PDA 197
Copying Only a Limited Number of References to PDA 199
What Happens to 'Figure' Type References in the PDA EndNote Library? 199

Working with the EndNote Library on PDA 201
Opening the Library 201
Opening a Reference for Viewing or Editing 202
Sorting References 202
Entering a New Reference 202
Editing a Reference 203
Deleting a Reference 204
Attaching a Note to a Reference 205
Searching References 205
Beaming References Between PDAs 207
Beaming a Reference or a Group of References 207
How Beaming Works 207
Viewing Statistics About the PDA EndNote Library 208
Customizing the PDA EndNote Library 208
Show Splash Screen at Startup 209
Default Reference Type 209
Scroll Button Behavior in Edit View 210

11. Using RefViz© with EndNote 211
What is RefViz? 211
How Does RefViz Work? 212
Technical Requirements for Installing RefViz 213
Getting and Installing RefViz 213
Installing RefViz 213
Activating RefViz 213
Working in RefViz 217
Starting RefViz 217
Working in RefViz 217
Exporting References from RefViz Back to EndNote 224

Appendix A: Online Resources to Learn More About EndNote 225

Appendix B: Online Resources to Help Writing for Publication 227

Appendix C: Downloading Files, Filters, and Styles for EndNote 229

Index 231

CHEAT SHEET FOR ENDNOTE

AN OVERVIEW OF USING ENDNOTE FOR REFERENCE MANAGEMENT

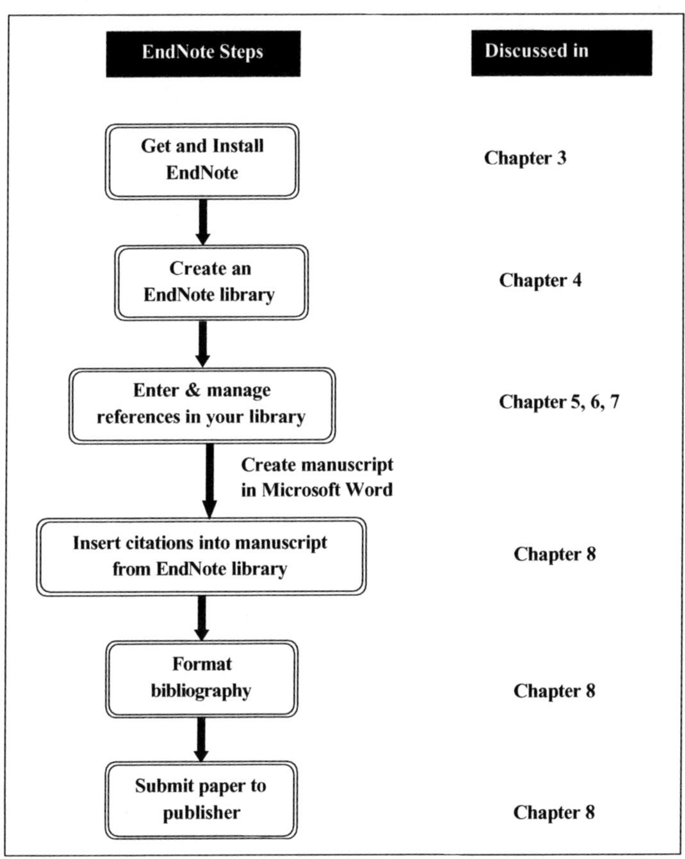

INPUT AND OUTPUT OPTIONS FOR ENDNOTE LIBRARY

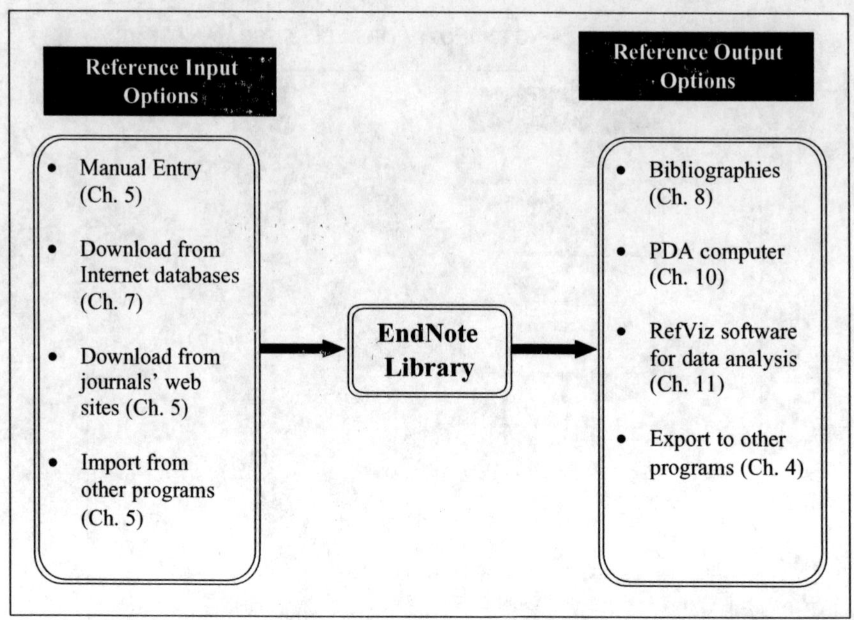

STEPS IN USING ENDNOTE WITH INTERNET DATABASES

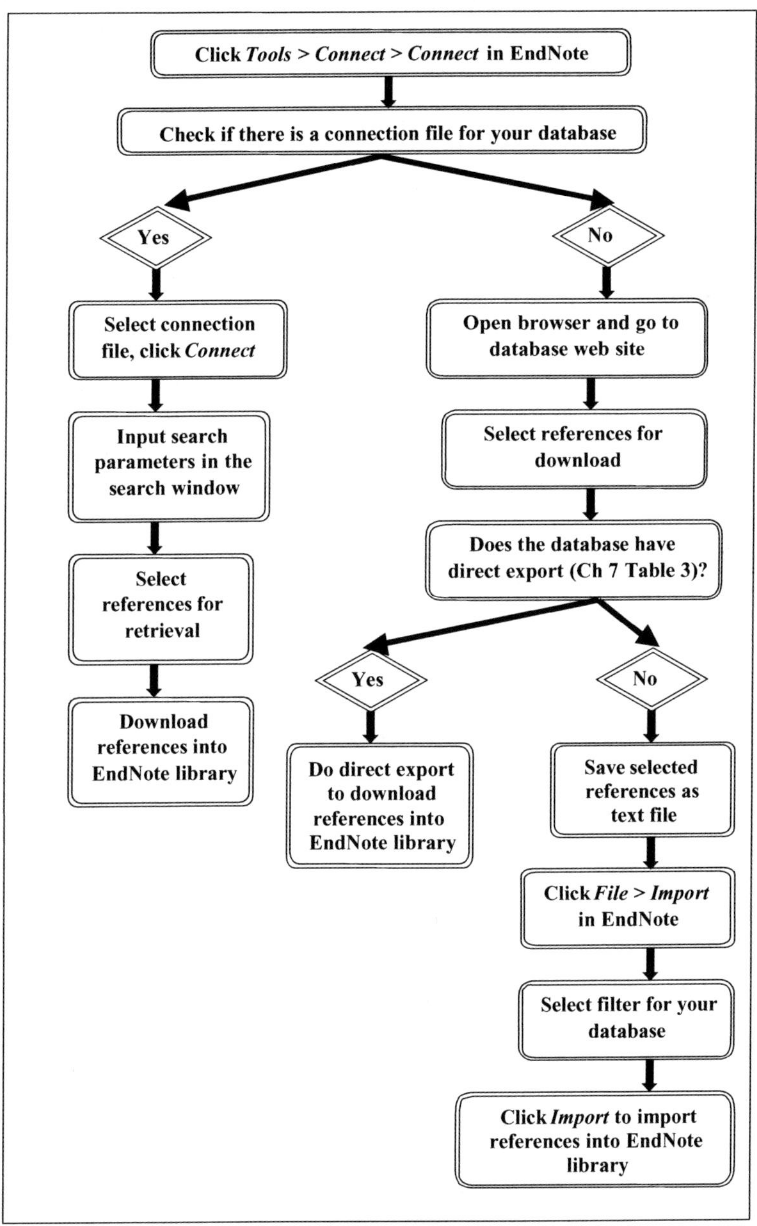

MAIN ENDNOTE WINDOW

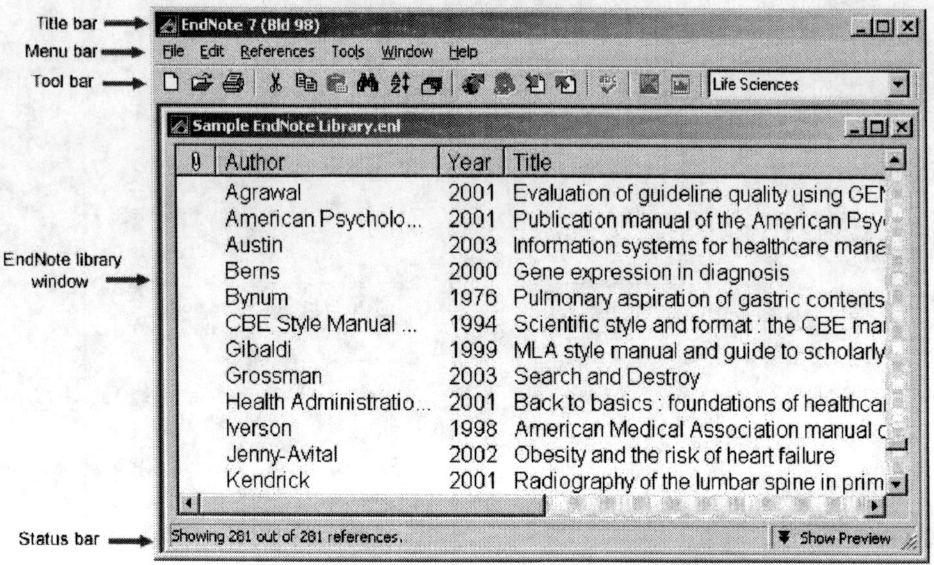

THE THREE TOOLBARS IN ENDNOTE

THE MAIN TOOLBAR

KEYBOARD SHORTCUTS

EndNote Keyboard shortcuts

EndNote Keyboard short cuts	
Keyboard short cuts help you work in EndNote with minimal use of mouse	
Key Command	**Function**
CTRL+N	Create a new reference
CTRL+CLICK	Select more than one reference
SHIFT+CLICK	Select a range of references
CTRL+E	Open selected reference(s)
CTRL+W	Close the active window
CTRL+SHIFT+W	Close all windows of the same type as the active window
TAB	Select the next field
SHIFT+TAB	Select the previous fields
CTRL+D	Delete a reference
CTRL + A	Select All / Unselect All references
CTRL + K	Copy Formatted references
CTRL + F	Search reference command
CTRL + Y	Launch a spell-checker (at least one reference must be open for this to work)
CTRL + R	Change text command
CTRL + H	Show all references
CTRL + G	Open link
CTRL + J	Find a word or phrase in the open reference

Keyboard shortcuts to access the main menu bar

Keyboard short cuts to access the menu bar	
Use these keyboard short cuts to access the main menu bar of EndNote and then use the down arrow key to access the submenu items. For example, by pressing Alt + F you will access the main File menu and then press down arrow key to access the submenu items such as New, Open, Close Library etc.	
Key Command	**Access to menu item**
ALT + F	File
ALT + E	Edit
ALT + R	Reference
ALT + L	Tools
ALT + W	Window
ALT + H	Help

Keyboard shortcuts to access CWYW

Keyboard Short Cuts to Access CWYW	
NOTE - you can create custom short cuts by setting CWYW preferences (see chapter 8). This is specially useful for commands for which there is no key defined by default such as format figure etc.	
Return to Word	ALT + 1
Find Citation(s)	ALT + 7
Go to EndNote	ALT + 1
Format Bibliography	ALT + 3
Insert Selected Citation(s)	ALT + 2
Edit Citation (s)	ALT + 6
Insert Note	ALT + 0
Edit Library Reference	ALT + 5
Unformat Citation	ALT + 4
Remove Field Codes	None
Export Traveling Library	ALT + 8
Find Figure	None
Generate Figure List	None
CWYW Preferences	ALT + 9
Help	None

Keyboard shortcuts for standard Windows commands

Keyboard shortcuts for Standard Windows Commands	
CTRL+S	Save
CTRL +P	Print
CTRL + Q	Exit a program
CTRl + X	Cut
CTRL + C	Copy
CTRL + V	Paste
CTRL + B	Turn bold on or off
CTRL + U	Turn underline on or off
CTRL + I	Turn italics on or off
CTRL + Z	Undo
CTRL + O	Open a file (library)
F1	Display the help file
CTRL + Esc or Windows	Open or close the Start menu
ALT + Tab	Switch back to a running program
Hold ALT, press Tab repeatedly	Switch to another running program
Windows + E	Open Windows Explorer to My Computer
Windows + D	Minimize open windows or (restore minimized windows)
ALT + F4	Close an open window
Shift + F10	Open the right click menu
CTRL + Home	Go to the beginning
CTRL + End	Go to the end

1 Introduction to Reference, Bibliography, and Citation

> **Things You will Learn in This Chapter**
>
> - Basic concepts about some essential terms such as 'reference,' 'reference list,' 'bibliography,' and 'citation' and how they relate to the anatomy of a manuscript.
> - Briefly, about referencing styles including the Harvard style and the Vancouver style.
> - The benefits of using information technology instead of a pile of index cards for effective reference management.

Research and writing are integral parts of the professional work of healthcare and biomedical professionals. Scientific manuscripts commonly include references to related information in literature. The inclusion of references in manuscripts substantiates arguments with evidence, as well as acknowledges the source of information being referred. References may be cited from a variety of sources such as journals, books, conference proceedings, magazines and newspapers, and the Internet. This chapter discusses the basic concepts related to the process of referencing as a foundation to the effective use of reference management software programs, such as EndNote.

Basic Concepts

Figure 1 outlines the concepts of reference, reference list, bibliography, and citation as they relate to the anatomy of a manuscript.

Reference

A reference is a short description or note that contains information about the source. Simply put, a reference is the 'address' of the source. References enable the reader to access and verify the original source of information; by knowing the address of

---

EndNote is a registered trademark of The Thomson Corporation.

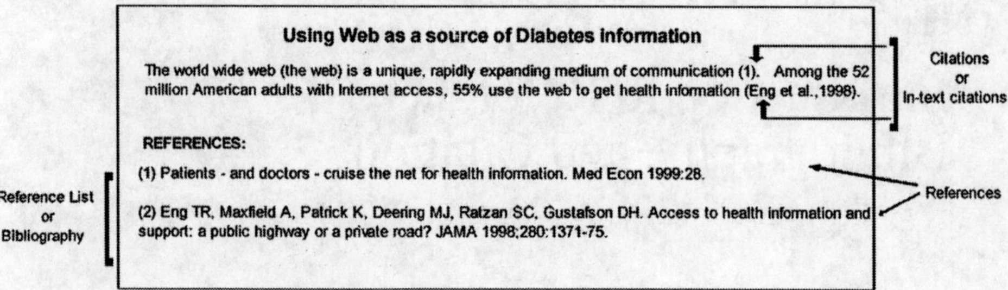

FIGURE 1. Anatomy of a manuscript (showing two different referencing styles just for illustration)

the source, a reader can look up the journal/book/web site, and so forth, in which the original material was published.

The following are some examples of references from commonly used sources:

Book Reference

Kohn Linda T., Corrigan Janet, Donaldson Molla S. To err is human: building a safer health system. National Academy Press, 2000.

Journal Article Reference

Jenny-Avital, E. R. Obesity and the risk of heart failure. N Engl J Med. 2002; 347(23); 1887–9.

Magazine Article Reference

Posner, M. I. (1993, October 29). Seeing the mind. Science, 262, 673–674.

Electronic Source Reference

Jonathan Amos. Scientists clone 30 human embryos. http://news.bbc.co.uk/2/hi/science/nature/3480921.stm Accessed 12 February, 2004.

Reference List and Bibliography

References may be included at the end of the manuscript, as endnotes, or included at the bottom of each page, as footnotes. A reference list is a numbered or alphabetically sorted list of references that are actually cited in the text of the manuscript as endnotes or footnotes. Bibliography is a term typically used to indicate a comprehensive list of all the resources the author has consulted during the course of the research. It may include resources in addition to those cited in the text. Note that the terms bibliography and reference list are often used interchangeably in common practice.

Citation

When references are included as endnotes/footnotes, how does the reader know which quote or text in the manuscript comes from which reference source? This is accomplished by putting a link to the reference in the body of the manuscript in a short form, called "citation," or "in-text citation."

> **Technical Tip:** Although I have described the 'puristic' concept of the terms reference, citation, and bibliography, be aware that these terms are frequently used interchangeably.

Common Referencing Styles

There are two main styles of formatting in-text citations and references in a document: the **author-date style** and the **footnote/endnote style** (also known as the **numbered style**). In addition there are a variety of other styles recommended by various journals and professional associations such as the Modern Language Association (MLA), the American Psychological Association (APA), and the American Medical Association (AMA).

Harvard style is an example of the author-date style. In this style, the in-text citation consists of the name of the author(s) and the year of publication, and an alphabetically sorted list of references is included at the end of the manuscript. The following is an example of the Harvard Style. Note that in the reference list, the two articles are listed in alphabetical order. The Berland article is listed first in the reference list though it is cited second in the text of the manuscript.

Harvard Style: Example

The world wide web is a unique, rapidly expanding medium of communication (Eng et al., 1998). Among the 52 million American adults with Internet access, 55% use the web to get health information (Berland et al., 2001).

Reference List:

Berland GK, Elliott MN, Morales LS, Algazy JI, Kravitz RL, Broder MS, et al. Health information on the internet: accessibility, quality, and readability in English and Spanish. JAMA 2001;285:2612–21.

Eng TR, Maxfield A, Patrick K, Deering MJ, Ratzan SC, Gustafson DH. Access to health information and support: a public highway or a private road? JAMA 1998;280:1371–75.

Vancouver style is an example of the footnote/endnote style. In this style, the in-text citation consists of a reference number, and a numbered reference list in order of appearance of the reference is included at the end of the manuscript. The following is an example of the Vancouver style. Note that in the reference list, the two articles are listed in numerical order. The Eng article precedes the Berland article because it is referenced first in the manuscript.

Vancouver Style: Example

The world wide web is a unique, rapidly expanding medium of communication (1). Among the 52 million American adults with Internet access, 55% use the web to get health information (2).

Reference List:

(1) Eng TR, Maxfield A, Patrick K, Deering MJ, Ratzan SC, Gustafson DH. Access to health information and support: a public highway or a private road? JAMA 1998;280:1371–75.

(2) Berland GK, Elliott MN, Morales LS, Algazy JI, Kravitz RL, Broder MS, et al. Health information on the internet: accessibility, quality, and readability in English and Spanish. JAMA 2001;285:2612–21.

Table 1 provides a list of common referencing styles used in various disciplines and resources to learn more about them.

TABLE 1. Common referencing styles

Style	Discipline	For more information about the style
American Psychological Association (APA) style	Psychology	American Psychological Association. *Publication Manual of the American Psychological Association.* Washington, DC: American Psychological Association, 2001. 5th ed.
American Medical Association (AMA) style	Health/ Medicine	Iverson C, American Medical Association. *American Medical Association Manual of Style: A Guide for Authors and Editors*. Baltimore, MD: Williams & Wilkins, 1998.
National Library of Medicine (NLM) style	Medicine	Patrias K. National Library of Medicine recommended formats for bibliographic citation. Supplement: Internet formats [Internet]. 2001.
American Institute of Physics (AIP) style	Sciences in general	American Institute of Physics. *AIP Style Manual.* 1990–1997. Available at http://www.aip.org.pubservs/style/4thed/toc.html
Council of Biology Editors (CBE) style	Biology	CBE Style Manual Committee. *Scientific Style and Format: The CBE Manual for Authors, Editors, and Publishers.* Cambridge; New York: Cambridge University Press, 1994.
Modern Language Association (MLA) style	Literature, arts, and humanities	Gibaldi J. *MLA Style Manual and Guide to Scholarly Publishing.* New York: Modern Language Association of America, 1999.

Using Information Technology for Effective Reference Management

Managing references and creating appropriately formatted bibliographies are time-consuming, error-prone, and cumbersome processes when performed manually using the traditional method of storing references in a pile of index cards. Some inherent problems with the manual method include:

(A) Scientific writing is an iterative process; references are frequently added, edited, or deleted while writing a manuscript. This requires changing in-text citations in the body of the manuscript, and ensuring that the citations in the text correspond to the appropriate references in the bibliography.

(B) Creating an appropriately formatted bibliography in compliance with the styling standards set by a publication requires considerable time and effort. For example, what would be an appropriate abbreviation for the journal name New England Journal of Medicine—NEJM or N Eng J Med? Should you use 'et al.' for more than three author names, or more than four, or never? How should you describe author names in a given bibliographic style required by a specific publication—should it be Jones, SK or Smith K Jones? These questions often baffle authors while writing manuscripts.

(C) If a manuscript has to be resubmitted to another publication that requires a different style of bibliography formatting, the process would require manually reformatting all the references into the new style, again, a time-consuming task.

Fortunately, new, sophisticated technology applications, called reference management software programs, are now available to facilitate the challenging process of reference management by allowing users to create personal digital libraries. Chapter 2 discusses various reference management software programs and their functions in details. Subsequent chapters in the book provide step-by-step instruction on using EndNote®, one of the most commonly used reference management programs.

2
Reference Management Software Programs

> **Things You will Learn in This Chapter**
>
> - What a reference management software program is.
> - Functions of reference management software programs.
> - About various reference management software programs and some criteria for choosing the right program.
> - About comparing EndNote®, ProCite®, and Reference Manager®.
> - About other reference management software programs, including freeware/shareware programs, and their features.

What is a Reference Management Software Program?

These are software programs that simplify the process of reference management by allowing the user to collect, store, and organize references, insert citations at the appropriate place in the body of the manuscript, and generate a list of the references in an appropriately formatted bibliographic style.

Reference management programs were first introduced in the 1980s and have been gaining popularity since. In a 2001 survey conducted by *Scientist* magazine, 76% of respondents indicated that they use reference management software to organize their reference material collection[1]. As seen in Figure 1, this survey indicated that EndNote was the most popular program among the 250 scientists polled.

Reference management programs work by creating a personal digital library (or, database) of references. The references in a digital library can be easily searched, sorted, and inserted in a manuscript to create accurate bibliographies.

---

[1]Perkel JM. The essential software toolbox. The Scientist 2001;15 (14):19

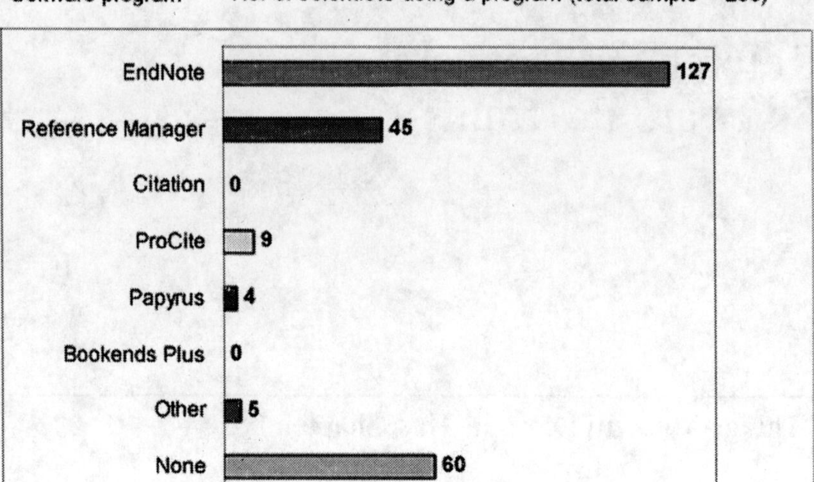

FIGURE 1. Reference management software program user survey

Functions of Reference Management Software Programs

(A) Storing and Managing References

By storing references in a digital library, these programs provide you functionality inconceivable in the analog world of paper index cards—for example, they can search the library by multiple criteria such as author, title, journal, year, or keywords. Digital libraries can be stored in a computer eliminating the need for shelf space for index cards. They can also be backed up easily to avoid losing reference data.

Some examples of the types of references that can be stored in a digital library include:

- Articles from journals, magazines, and newspapers;
- Books and book sections;
- Grants, theses and reports;
- Figures;
- Charts, tables and equations.

A reference needs to be entered into a digital library only once, either by manually typing it in the library (Chapter 5) or automatically downloading it from an Internet database (Chapter 7). Then it can be used as many times as needed to create bibliographies.

(B) Creating Bibliographies

You can easily insert in-text citations into the body of the manuscript and automatically create an appropriately formatted bibliography using a reference management software program. The useful functions of these programs for creating bibliographies include:

- References are frequently added, edited, or deleted during the iterative process of writing. These programs greatly facilitate the process of automatically changing citation numbers in the body of the manuscript and ensuring that the in-text citations correspond to the bibliography.
- The bibliography can be automatically formatted according to a specific style such as the Vancouver style or the APA 5th style, or the style conforming to the requirements of the publisher.
- These programs eliminate the guess work out of accurately formatting bibliography and give you correct answers to questions such as 'should the New England Journal of Medicine be cited as NEJM or N Eng J Med?' and 'should et al. by used for more than three author names, or more than four, or never?'
- If you need to submit your paper to another publication for reconsideration, these programs will save you trouble by easily modifying the in-text citations and reference list according the style specification of the new publication.

Chapter 8 describes using EndNote to create bibliographies.

(C) Searching and Retrieving References from Online Databases

You can easily search various online databases such as PubMed®, Ovid®, the Library of Congress, and various university catalogs using these programs (Chapter 7). This could minimize the effort of going to the websites of these Internet databases and learning their search mechanisms and syntax. Furthermore, these programs allow you to automatically retrieve selected references in your digital library without the need for manually entering references by typing. Automatic entry of references provides the following benefits:

- Typing errors, such as in authors' names or titles, are eliminated.
- Correct abbreviations of journal names are entered in the library.
- In addition to the essential reference information such as title, author(s) names, journal, and date of publication, extra information about a given reference is downloaded into the library, such as the abstract and the URL.

(D) Working with a Handheld Computer

Digital libraries can be copied onto a handheld device, such as a Palm computer for portability and for sharing information with colleagues through the infrared beaming technology (Chapter 10).

Various Reference Management Software Programs

The various of reference management software programs available these days differ in their emphasis on ease-of-use vs. functionality. As the functionality of a software program increases, generally the learning curve for the program becomes steeper.

The three popular reference management software programs are EndNote®, ProCite®, and Reference Manager®, all offered by the California-based company ISI ResearchSoft (http://www.isiresearchsoft.com). These three software products have been around since the early 1980s, when they were three separate products from different companies.

Table 1 provides a detailed comparison between these three software programs. In summary, EndNote is one of the most popular reference management software

TABLE 1. Comparison of EndNote, Reference Manager, and ProCite

Feature	EndNote®	Reference Manager®	ProCite®
Version	7	10	5
Highlight	The most popular and easiest to use	Network capabilities with simultaneous read/write access	Flexibility to group references and create subject bibliographies
Search the Internet	Yes	Yes	Yes
Organize references	Yes	Yes	Yes
Organize images and files	Yes	No	No
Format bibliographies	Yes	Yes	Yes
Operating system	Mac and Windows	Windows	Mac and Windows
Reference size	64K character maximum	Unlimited	Unlimited
Maximum number of references	32,000	Unlimited	Unlimited
Maximum number of fields	40	37	45
Maximum number of reference types	28	35	50 supplied, can add
Create lists of images and files	Yes	No	No
Subject bibliography	Yes	No	Yes
Spell check	Yes	Yes	No
True network capabilities	No	Yes	No
Create a list of "favorite" styles	Yes	No	No
Reference grouping	No	No	Yes
Advanced searching capabilities	No	No	Yes
Search across multiple databases	No	Yes	No
Construct document with Microsoft Word templates	Yes	No	No
Number of preloaded output styles	1,000+	700+	620+

Adapted from http://thomsonisiresearchsoft.com/compare/ (May 2004).

TABLE 2. Other commonly used reference management software programs

Name	Vendor	URL	Platform	Price*
Library Master	Balboa Software	http://www.balboa-software.com/libinfo.html	Windows, DOS	$249.95
Bibliographix	Bibliographix GbR	http://www.bibliographix.com	Windows, DOS	Free basic version; 100 EUR for Pro version
Biblioscape	Biblioscape	http://www.biblioscape.com	Windows	$79–$699
BiblioExpress			Windows	Free
BiblioWeb Server			Windows; Web-enabled program	$99–$499
Biblio Palm			Palm OS 3,4,5	$49
Citation	askSam Systems	http://www.citationonline.net	Windows	$229.95–$299
Nota Bene	Nota Bene	http://www.notabene.com	Windows	$399–$449
GetARef	DatAid AB	http://www.getaref.com	Windows	$230
SquareNote	SQN Inc.	http://sqn.com/sqn5.html	Windows	Free to $49
RefWorks	RefWorks	http://www.refworks.com	Web-based system; Windows, Mac, Unix	$70 per year
WriteNote	ISI ResearchSoft	http://www.writenote.com	Web-based service; Windows, Mac, Linux	Information not available on the web site (July 18, 2004)
Bookends	Sonny Software	http://www.sonnysoftware.com	Mac	$99

*These are standard prices available on the web sites of these vendors. Check their web site for other pricing such as academic or group discount.

programs. It is an ideal tool for individual researcher and is available in student and standard editions. Reference Manager is a powerful bibliographic software, and is an ideal tool for large workgroups, networks, and collaborative projects. Reference databases can be shared over an intranet using a network edition of Reference Manager® for simultaneous read-write access by multiple users. ProCite is more sophisticated, and is usually used for special collections and catalogs. Unlike EndNote, it does not have a limit on the size of the reference library or the number of references in the library.

Table 2 provides a list of other commonly used reference management software programs. In addition to these commercially available programs, many other programs are available for free download on the Internet. None of these freely downloadable programs offer the features of a full bibliographic management tool; nonetheless you may find them useful for your professional needs. Table 3 provides a list of freeware (or shareware) reference management programs.

TABLE 3. Freeware/shareware reference management programs

Name	Vendor	URL	Platform	Description
Biblio-Express	Biblioscape	http://www.biblioscape.com/biblioexpress.htm	Windows	The freeware edition of Biblioscape with limited functionality
Papyrus	Research Software Design	http://www.researchsoftwaredesign.com	DOS, Windows, Mac	A full-featured software with ability to create bibliographies and search the Internet databases
Scholar's Aid 2000 Lite	Scholar's Aid	http://www.scholarsaid.com	Windows	Free software that helps download references from Internet databases. It can also be used to convert EndNote databases into a Scholar's Aid
Online Bibliography Builder	The University of Toronto	http://www.ecf.toronto.edu/~writing/bb.html	Web based; requires a browser with Javascript and frames capability	This software helps the researcher format the references in two styles only, the Chicago Manual of Style (author-date style documentation) and the IEEE style documentation (a reference-sequence system). Each bibliography entry can be generated by filling in an online form and clicking the submit button. The reference that is generated can then be pasted into the word processor. There is no functionality to save the references or provision to process a batch job
My Notes	Tim Takacs	http://www.tn-elderlaw.com/mynotes.html	Windows; Microsoft Access	A Microsoft Access Reference Manager. New references are inserted into the database by filling in the access form. The software offers functionality of searching for references within the database and printing reports. It does not however offer functionality like integration of the reference manager with Microsoft Word and automatic download of references from other Internet reference database like Medline/Ovid. It does not provide a choice of output styles

Which Program Should You Choose?

My suggestion is that you test-drive a software program before buying. Many vendors (including ISI ResearchSoft for EndNote) let you download a fully functional trial version of the program for free. Some of the important considerations for choosing the right software are:

(a) ***Individual Use vs. Network Installation.*** If you are buying software for individual use, you have a wide variety of choices. If you want to install a reference management program on the network of an organization, only a few software programs (such as Reference Manager) are suitable for network installation. EndNote provides only limited network capabilities.

(b) ***Platform.*** Keep in mind that most of the available programs run on Windows-based computers. If your computer has a different operating system such as Mac, you need to look carefully in the installation requirement of the software to make sure it can run on a Macintosh computer. Your other option is to use a web-based program as they require only a compliant browser (and not necessarily a compliant operating system).

(c) ***Ease of Use vs. Functionality.*** Ease of use often inversely correlates with functionality. Generally, programs that have many advanced features are somewhat complicated and cumbersome to use. EndNote is a popular software because it provides a happy medium for ease of use and functionality.

(d) ***Price.*** I have listed prices for some programs in Table 2 of this chapter and Table 1 of Chapter 3. Please refer to vendors' websites for more detailed information.

The rest of this book deals with learning how to use EndNote efficiently and effectively. I decided to focus on EndNote as the software of choice because it is the most commonly used reference management software among healthcare and biomedical professionals.

3
Getting Started with EndNote

Things You will Learn in This Chapter

- An overview of steps in working with EndNote.
- Technical requirements your computer should meet for installing and using EndNote.
- How to get and install EndNote on your computer.
- How to check your EndNote installation.
- How to fix the problem of EndNote menu not working in Microsoft Word.
- How to automatically update EndNote.
- How to uninstall EndNote.

An Overview of Working with EndNote

EndNote is one of the most popular reference management programs among biomedical and healthcare professionals. Figure 1 illustrates the series of steps commonly used in working with EndNote for scientific manuscript writing. This chapter discusses the first step in the workflow—getting and installing EndNote. If you have already successfully installed EndNote, you may ignore this chapter.

I have used version 7.0 of EndNote for Windows in this book to illustrate various tasks and examples. Although many of the concepts are similar, some examples may not work with earlier versions of EndNote.

Technical Requirements for Using EndNote

As a first step, make sure that your computer meets the following technical requirements for installing and using EndNote 7.0 for Windows.

Hardware Requirements

- An IBM PC or compatible computer with a Pentium or compatible processor (or higher);

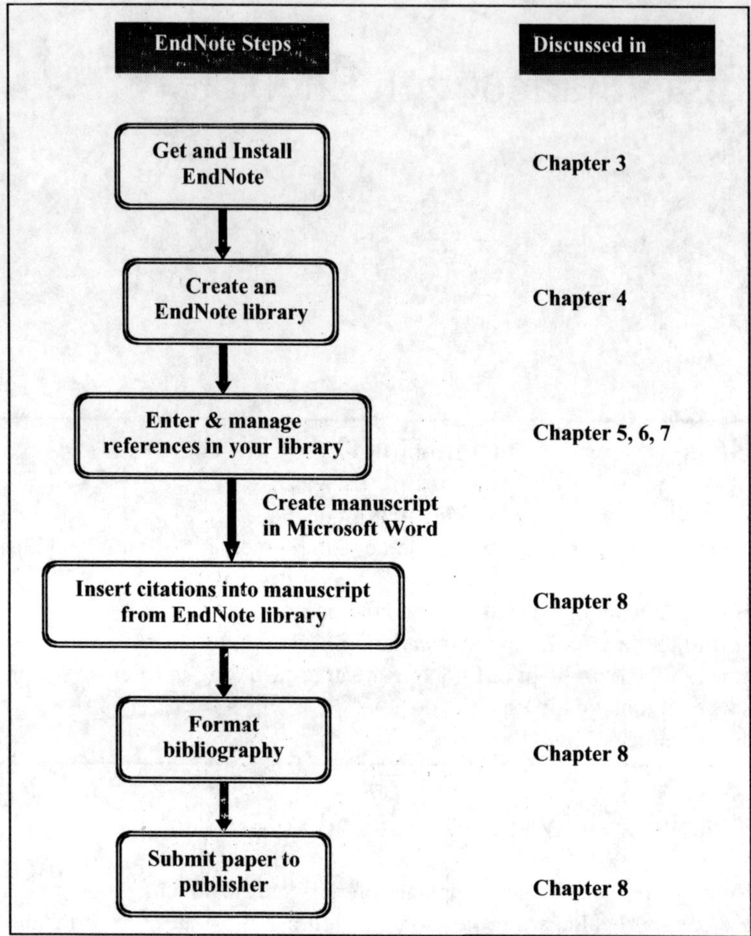

FIGURE 1. An overview of steps used in working with EndNote

- 64 MB of RAM or higher;
- A hard drive with at least 60 MB of free space;
- A CD-ROM drive as EndNote is distributed on a CD. If you download the software from the Internet, you don't need a CD-ROM drive in your computer;
- Internet connection if you plan to use EndNote to search Internet databases.

Note that most computers in use these days will meet these requirements.

Operating System (OS) Requirements

EndNote 7.0 runs under the following operating systems:

- Windows 2000;
- Windows ME;
- Windows XP.

In this book I discuss using EndNote on Windows OS based computers only. However, EndNote is available for Mac OS as well. Please note that the files created in the same version of EndNote are compatible across Windows and Mac platforms.

> **Technical Tip:** EndNote 7 also works with Windows 98SE and NT4. Except for the ability to use the EndNote for Palm application to transfer EndNote library to a hand-held computer, all other features of EndNote 7 are fully functional with these earlier versions of Windows.

Word Processor Compatibility

EndNote 7 is compatible with the following programs and formats:

- Microsoft Word 97/2000/XP (2002);
- WordPerfect 9 (2000) and 10 (2002);
- 'RTF (Rich Text Format)' files created with most word processors including; OpenOffice, StarOffice, and FrameMaker.

When you install EndNote, a submenu of EndNote functions is created in the word processor window by using the Cite While You Write (CWYW) function for Word and the EndNote Add-in function for WordPerfect. This allows you to access EndNote functions from your word processor screen without having to switch to EndNote. See Chapter 8 for more information about CWYW.

Hand-Held Computer Requirements

If you plan to use EndNote on a hand-held computer (also called Portable Digital Assistant or PDA), you need to make sure that your PDA meets the following requirements:

- It must be a Palm® OS based PDA with OS version 3.1 or higher;
- You will need a Serial or USB port on your desktop computer for HotSync operation;
- You will need a HotSync cradle or cable (generally included with the PDA);
- At least 4 MB RAM on the PDA.

At present, EndNote for Palm application is compatible with Palm OS-based devices only and not with PocketPC-type devices such as those manufactured by Compaq or Hewlett-Packard. The EndNote vendor's web site (www.endnote.com) does not contain any information whether they plan to make this available for the PocketPCs sometime in the future.

Another hint of caution: The EndNote vendor also advises that EndNote for Palm was designed to run on PDAs manufactured by Palm Inc., only. It may be possible to install and run it on PDAs with Palm OS manufactured by other vendors (e.g. Handspring or Sony), but EndNote manufacturer does not recommend or support doing so.

TABLE 1. Pricing for EndNote

EndNote 7 CD + Print Manual	$299.95
EndNote 7 download from the Internet	$239.95
Upgrade EndNote 7 CD + Print Manual	$99.95
Upgrade EndNote 7 download from the Internet	$89.95

Source: http://www.endnote.com; accessed May 15, 2004

Some example of EndNote compatible PDAs manufactured by Palm, Inc. are:

- All models from the Zire series;
- All models from the Tungsten series;
- m series: m125, m130, m500, m505, m515;
- Palm Vx.

Personally, I have used the EndNote for Palm application on a Palm IIIxe model without a problem. Chapter 10 describes using EndNote for PDAs.

Getting EndNote

You can buy the packaged software from the vendor or download it from the Internet. If you buy the packaged software, it comes on a CD-ROM and is accompanied by a paper manual. If you download it from the Internet, you receive an electronic help manual in PDF format.

Generally Internet download is cheaper than buying the packaged CD software. If you are using an earlier version of EndNote, it is cheaper to upgrade to version 7.0 than to buy a new product. Table 1 lists the pricing listed on the EndNote web site for EndNote 7 for a Windows product at the time of this writing; visit http://www.endnote.com to get latest prices.

Technical Tip: What is PDF? Portable Document Format, or PDF, is a file format developed to preserve the fonts, images, graphics, and layout of a document. When you share computer files such as Microsoft Word document with other people or on the Internet, a common problem is that the graphics and layout of the file are not displayed correctly on other computers. Using PDF format to share and distribute files ensures the integrity of the formatting and structure of documents. PDF was developed by Adobe Systems and PDF files can be viewed by anyone by using the free Adobe Acrobat Reader software. The software can be downloaded from the web site of the Adobe Systems at http://www.adobe.com/products/acrobat/readstep2.html. You can easily convert a variety of files such as Microsoft Word into PDF. While the software to view the PDF files is free, to create PDF files you need the full version of the Adobe Acrobat software, which is available at http://www.adobe.com/products/acrobatpro/main.html and is not free.

Trial Version of EndNote

You can also download a free 30-day trial version of EndNote at http://www.endnote.com/endemo.asp. The trial version has the full functionality of EndNote and is available for Windows as well as Macintosh OS.

Installing EndNote

You will have one of the following three installation scenarios:

A. Installing EndNote for the first time;
B. Upgrading EndNote from a previous version;
C. Installing EndNote onto a network.

> **Technical Tip:** In order for EndNote to work correctly with a word processor, a word processing system such as Microsoft Word or WordPerfect must be installed on your computer prior to installing EndNote.

A. Installing EndNote for the First Time

- Make sure you are not running any other programs on the computer.
- Insert the EndNote CD into the CD-ROM drive or click on the file you downloaded from the Internet as the EndNote program. The EndNote installation program will start.
- Follow the instructions on the screen to complete the installation.

> **Alert:** Sometimes the EndNote installer does not start after inserting the CD. This generally happens if you don't have the 'Autoplay' feature for CDs enabled. If so, you can easily start EndNote installation by performing the following steps:
>
> - Click on the *Start* button on your desktop.
> - Click *Run* (Figure 2).
> - In the next screen, type D:\setup, where D is the drive letter for the CD-ROM drive on your computer (Figure 3).
> - Click *OK* or Press *Enter.*

B. Upgrading from an Earlier Version of EndNote

The installation procedure remains the same as above. You do **NOT** need to uninstall the previous version from the computer. If during the installation a previous version of EndNote is found on the computer, you will see the screen in Figure 4.

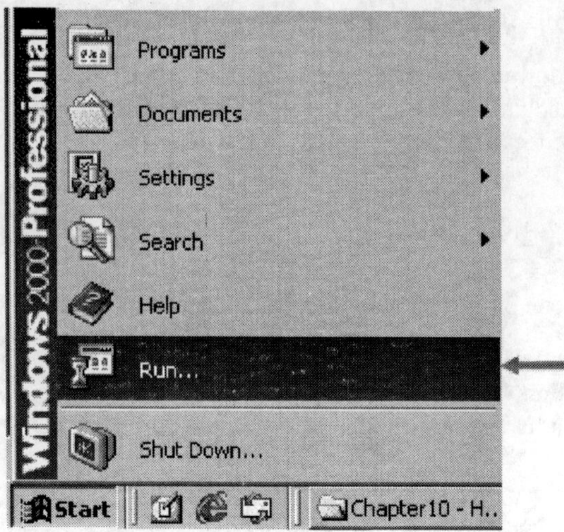

FIGURE 2. Selecting Start > Run

- Click on one of the radio buttons to backup or overwrite your older files. I recommend backing up the files, as you may need them later.
- Click *Next* and follow instructions to continue with the installation.

If you choose to backup files, they are placed in a folder called "Backup"—generally the path to this folder is C:\Program Files\EndNote\Backup if you used the default options during the installation.

> **Alert:** Prior to upgrading to the new version, remember to unformat any papers you may later need to alter/reformat using the new version. Reformatting with the new version may result in a missing or extra reference, which creates inaccuracies in your paper. If you unformat the paper and reformat anew with the new version, this problem can be easily avoided. See Chapter 8 for more information about how to unformat and reformat your paper.

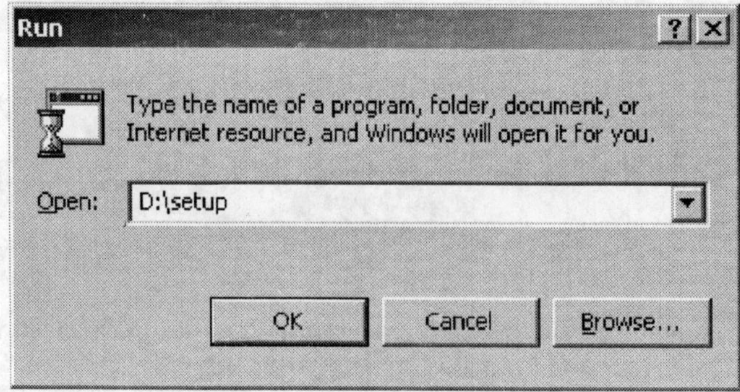

FIGURE 3. Typing in the "Run" box

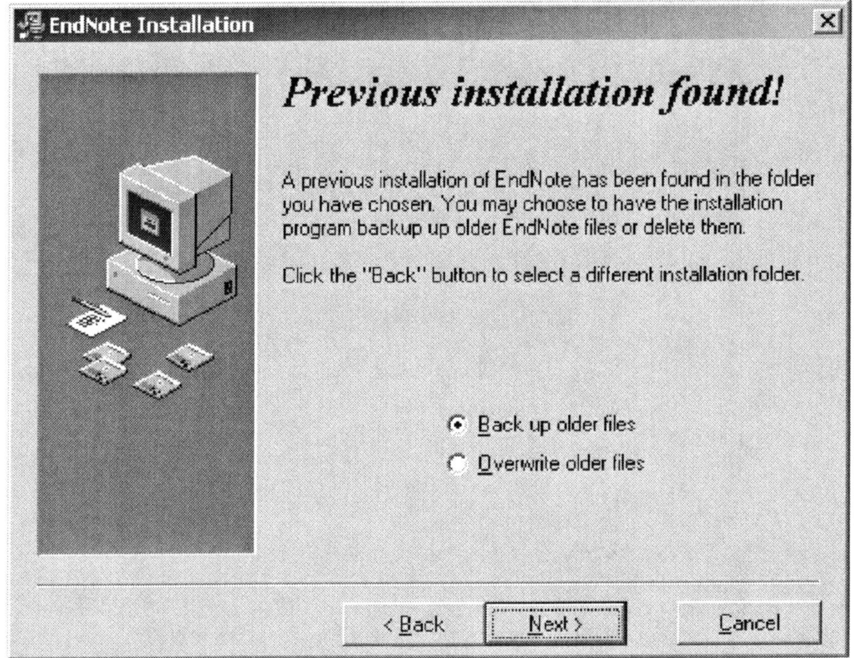

FIGURE 4. Selecting Backup of files

Technical Tip: If you are upgrading from an older version, you will need the serial number on the original EndNote CD or in the manual. Not surprisingly, many people lose this number. A simple way to retrieve this number is:

- Click *Help > About EndNote* in the EndNote menu (Figure 5).
- You should see the EndNote serial number in the next flash screen (Figure 6). Note that I have done a 'white-out' of the serial number of my copy of EndNote.

C. Installing EndNote onto a Network

According to its manufacturer, even though EndNote can be used across a network, it was not designed with specific networking capabilities in mind. EndNote can be installed on a server to allow multiple user access to EndNote libraries or the EndNote program. Generally this book does not deal with network installations, as they require a more skilled system administrator. You should seek technical support at your organization to install EndNote on a network.

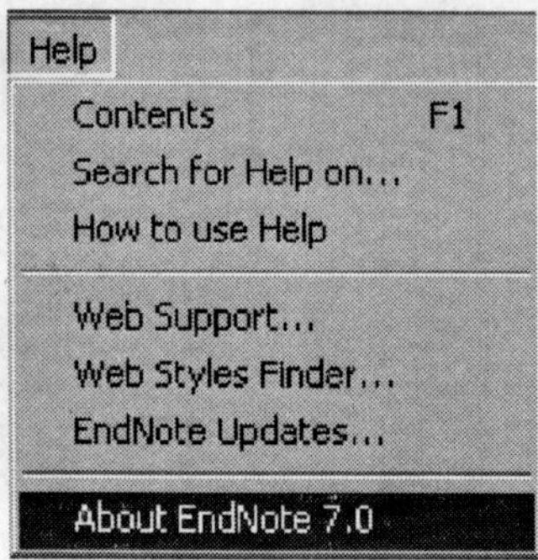

FIGURE 5. Selecting About EndNote menu

Checking EndNote Installation

Checking EndNote

At the end of installation, you should verify that EndNote is installed correctly on your computer. To do this:

- Click *Start* > *Programs*. You should see EndNote as one of the programs in the list.
- When you click on EndNote, you will see various options under EndNote. Click on *EndNote program* (Figure 7) and it should start EndNote.

If you can successfully do these steps, your EndNote installation is complete.

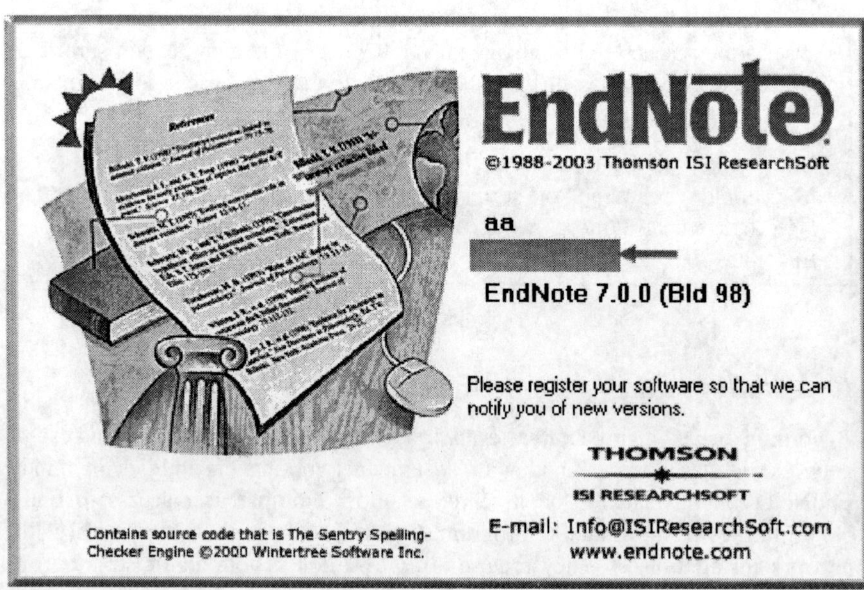

FIGURE 6. About EndNote screen

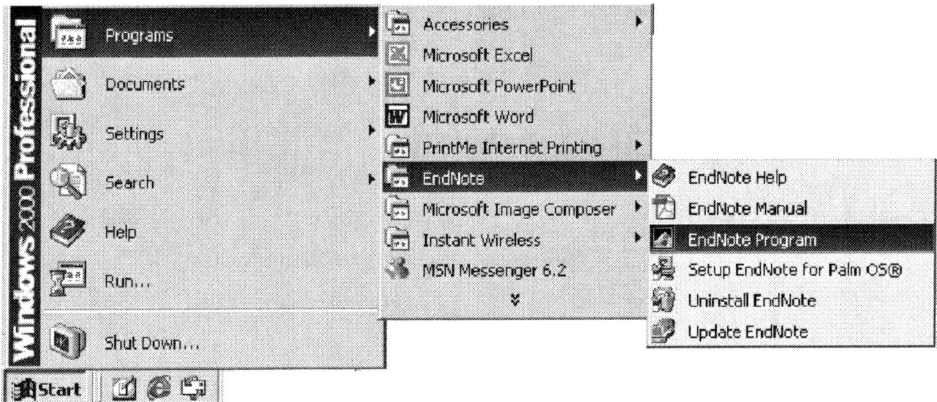

FIGURE 7. Checking EndNote installation

Checking Word Processor Support

If Microsoft Word was available on your computer before you installed EndNote (most likely it was, as most computer these days are shipped with a word processor installed), appropriate files to support EndNote functions within Word, that is, CWYW, should automatically be installed.

To check whether these functions have been correctly installed, perform the following steps:

- Start Microsoft Word.
- Click on the *Tools* menu.

You should see EndNote 7 under the Tools menu along with its submenu and commands (Figure 8).

What to do if you upgrade your version of Word after installing EndNote?

You should run the EndNote setup again to ensure correct installation of the menu for Word. Perform the following steps:

- Insert the EndNote CD.
- After the next few screens, select the radio button with *Custom installation* option (Figure 9). Click *Next.*
- In the next screen, make sure only the *Add-in Support* option is checked and all other boxes are unchecked. Click the *Next* button (Figure 10).

This should install the proper support for your version of the Microsoft Word.

What to do if EndNote commands do not appear in Microsoft Word?

This is one of the most common problems encountered during the EndNote installation and is particularly common when installing EndNote on an office computer.

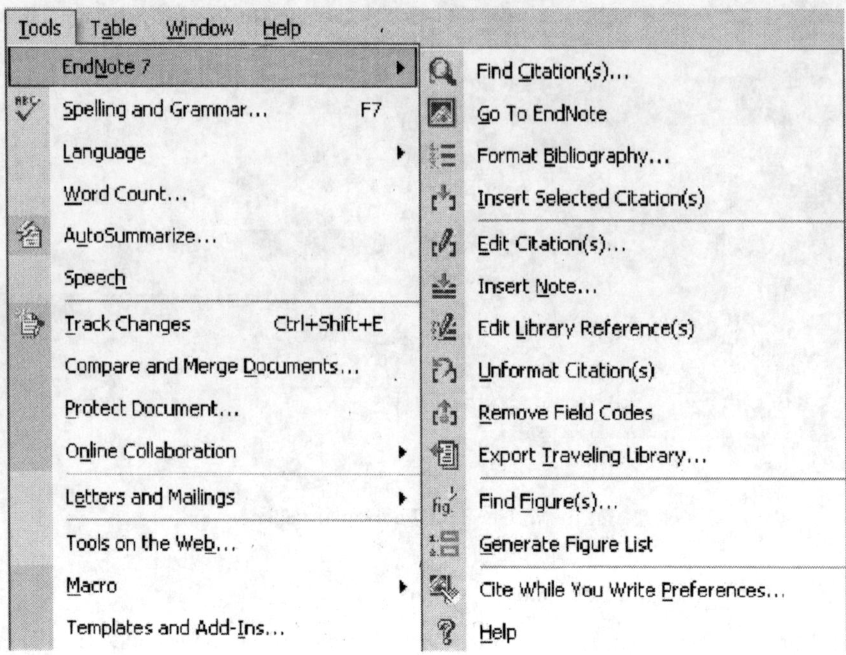

FIGURE 8. EndNote submenu in Microsoft Word

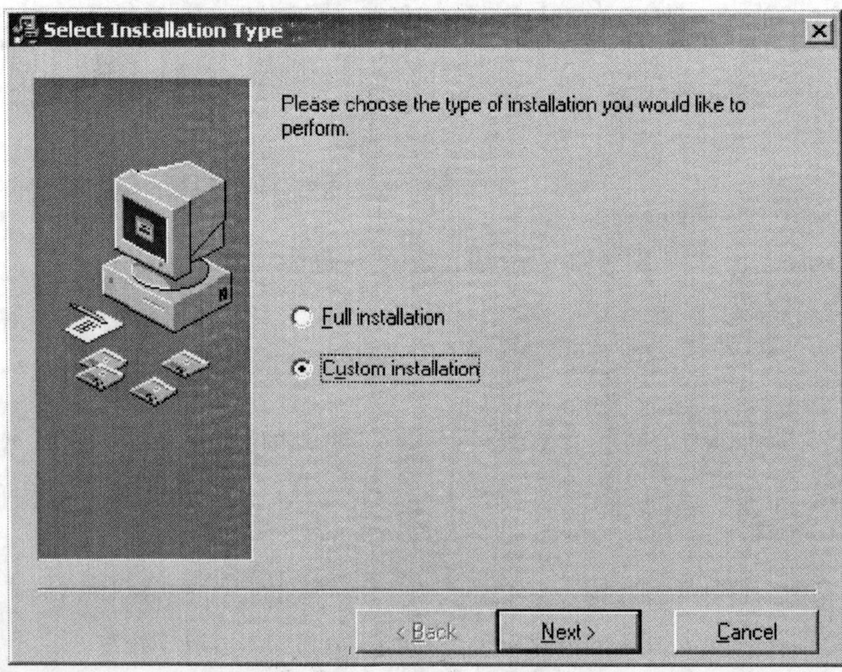

FIGURE 9. Selecting Custom Installation

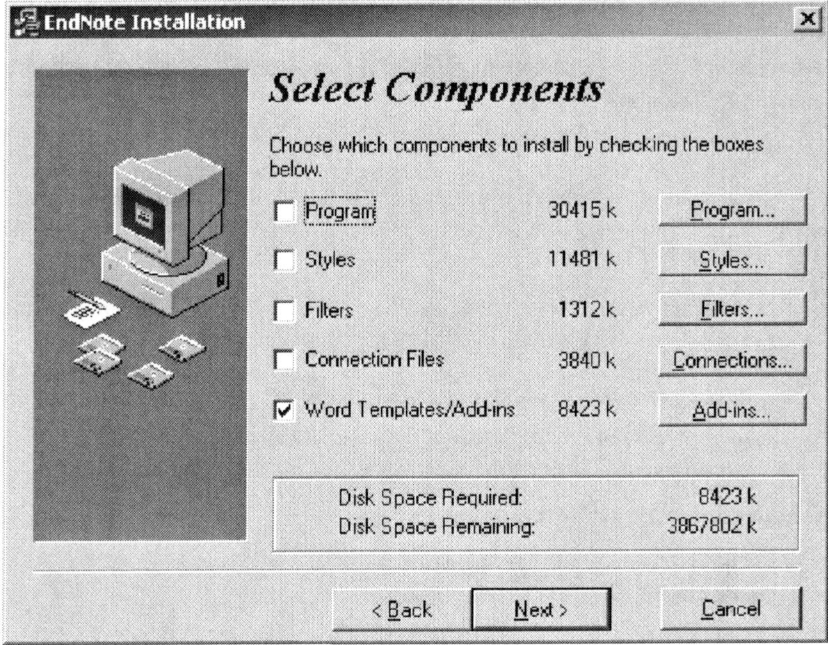

FIGURE 10. Selecting the Add-in Suppport option

Many users in a networked organization have restricted privileges and no administrative rights on their computer. Because of this, EndNote cannot properly access the Windows system, which causes difficulty in integrating installation of the EndNote menu in Word.

It would be helpful for you to understand the fundamentals of how CWYW works to troubleshoot this problem. During installation, EndNote automatically places two special files in the Startup Folder of the Microsoft Word program. Table 2 lists the CWYW files needed for various versions of EndNote and Word. These files are responsible for the appearance of the EndNote menu under Microsoft Word. If EndNote is unable to place these files in this Startup folder, the EndNote menu will not appear in Word. When you have restricted privileges on a

TABLE 2. CWYW files for various versions of EndNote and Word

EndNote version	Word version	Files
5	97/2000/XP/2003	EndNote5.cwyw.dot EndNote5.cwyw.wll
6	97/2000/XP/2003	EN6CWYW.dot EN6Cwyw.wll
7	97/2000	EN7cwyw.dot EN7cwyw.wll
7	XP/2003	EN7cwyw.dot EN7cwyw.wordxp.wll

computer, the Startup folder may not be accessible to you and, therefore, during the installation process EndNote is unable to copy these two files in the Startup folder.

If you don't see the EndNote menu in Word after EndNote installation, try the following solutions to solve this problem:

Solution 1: Run the EndNote installation again.

Try this solution first as this is the simplest solution. By running the installation again, may be EndNote will be able to locate the Startup folder and copy these files.

- Insert the EndNote CD.
- After the next few screens, select the radio button with *Custom installation* option (Figure 9).
- In the next screen, make sure only the *Add-in Support* option is checked and all other boxes are unchecked. Click the *Next* button (Figure 10).
- Once the installation is complete, open EndNote and a library.
- Then open Word and check for EndNote commands under the Tools menu.

Solution 2: Install Add-ins manually.

If solution 1 does not solve the problem, you may need to copy these files manually from EndNote to the Startup folder. To do this:

A. Find the location and path of the Startup folder.

- In Word, click *Tools > Options* (Figure 11).
- Click the *File Locations* tab.
- Highlight the Startup folder by clicking on it. Click *Modify* (Figure 12).
- In the next dialog box, click on the "Look In" dropdown at the top of the window and write down the path to the Startup folder. For example, in Figure 13, the path to the Startup folder is C:\Documents and Settings\Administrator\Application Data\Microsoft\Word\Startup. Click *Cancel*. Click *Cancel* in the Options window as well.
- Close Word and any other applications such as Outlook.

B. Then, locate the EndNote folder.

- Launch Windows Explorer (Figure 14) by clicking *Start > Programs > Accessories > Windows Explorer* and locate your EndNote folder. If you accepted

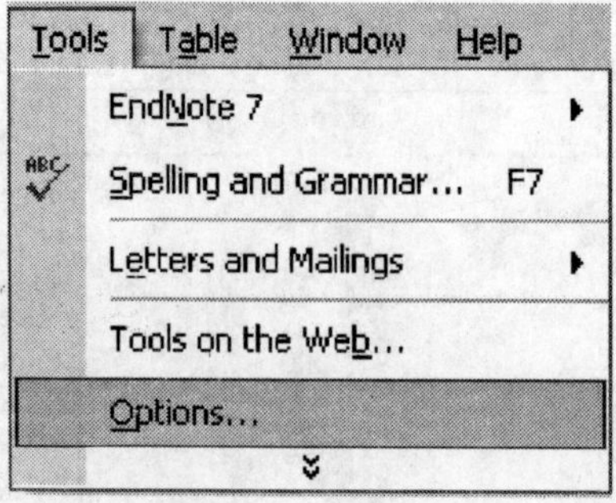

FIGURE 11. Selecting Tools > Options in Microsoft Word

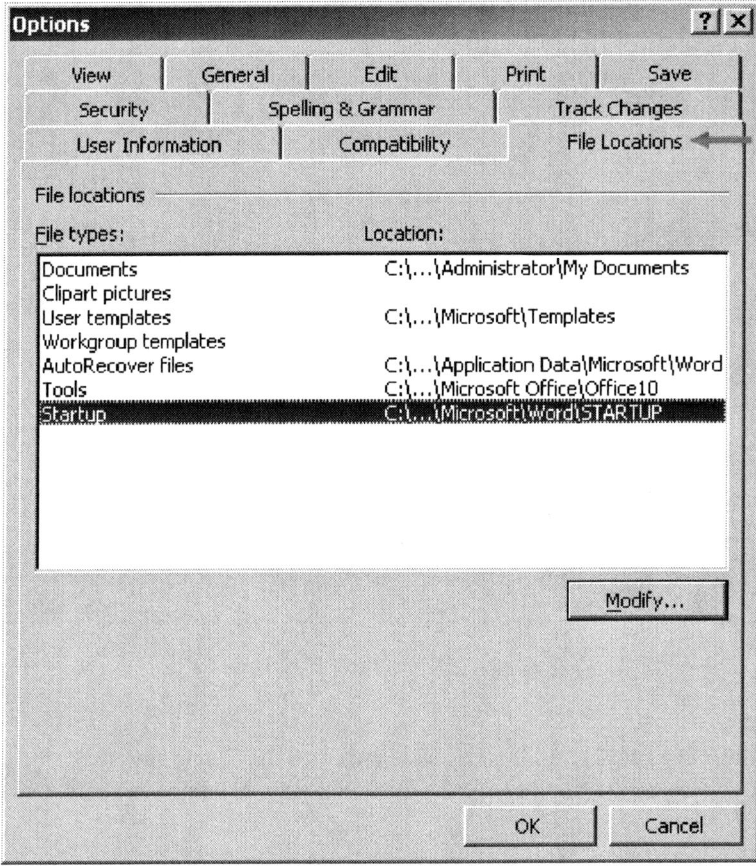

FIGURE 12. The Startup folder in the File Locations tab

the default options during the installation, the path for this folder should be "C:\Program Files\EndNote".

- There are two files in this folder you need to copy, depending on which version of EndNote and Word you are using (Table 2).
- Select these two files (by clicking on them while holding the *CTRL* key) and click *Edit* > *Copy*.

Technical Tip: If you do not see the .wll and .dot file extensions in the Windows Explorer:

- Click *Tools* > *Folder Options* in Windows Explorer (Figure 15).
- Select the View tab and **uncheck** the box next to "Hide file extensions for known file types" (Figure 16).
- Click *OK*. You should now see the file extensions.

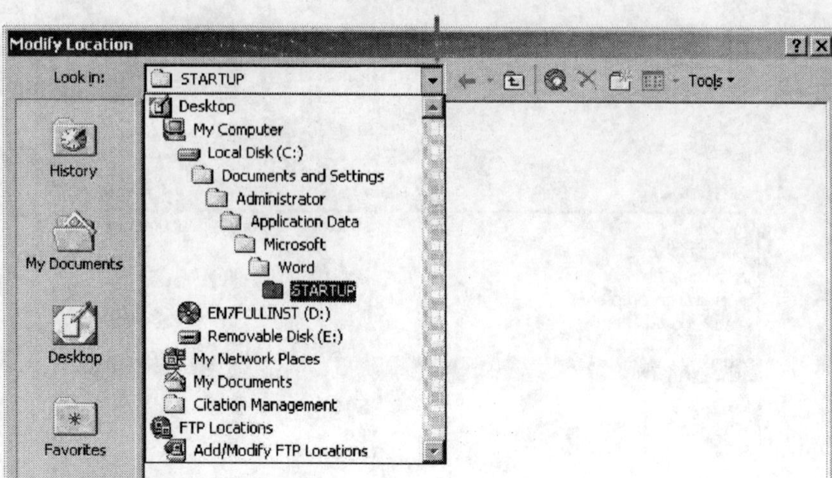

FIGURE 13. Locating the Startup folder

C. Then copy files from the EndNote folder to the Startup folder.

- Locate the Startup folder using Windows Explorer using the path you noted in step A.
- Paste the above two files (from Step B) in the Startup folder by clicking *Copy > Paste*.

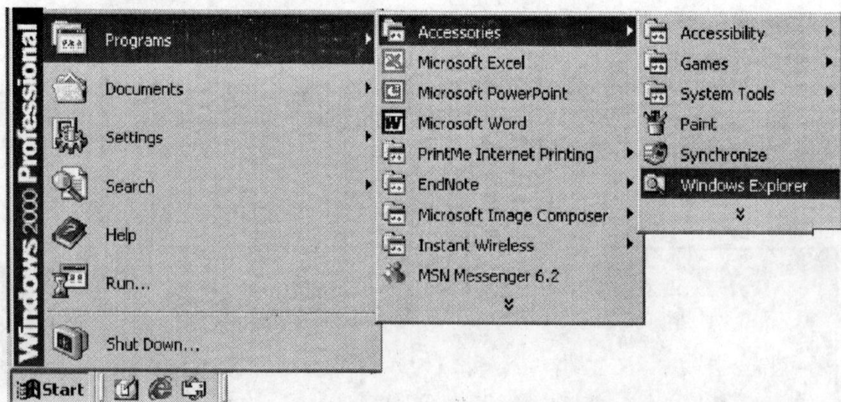

FIGURE 14. Launching the Windows Explorer

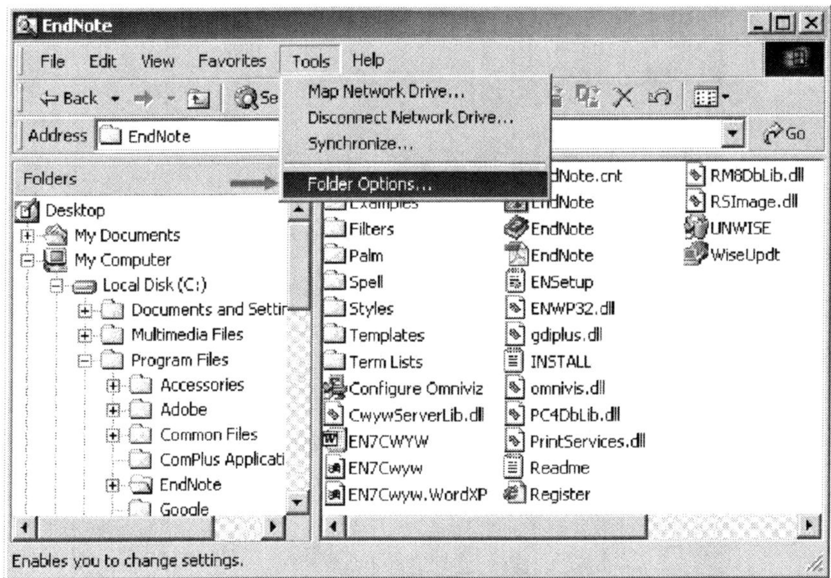

FIGURE 15. Selecting Folder Options in Windows Explorer

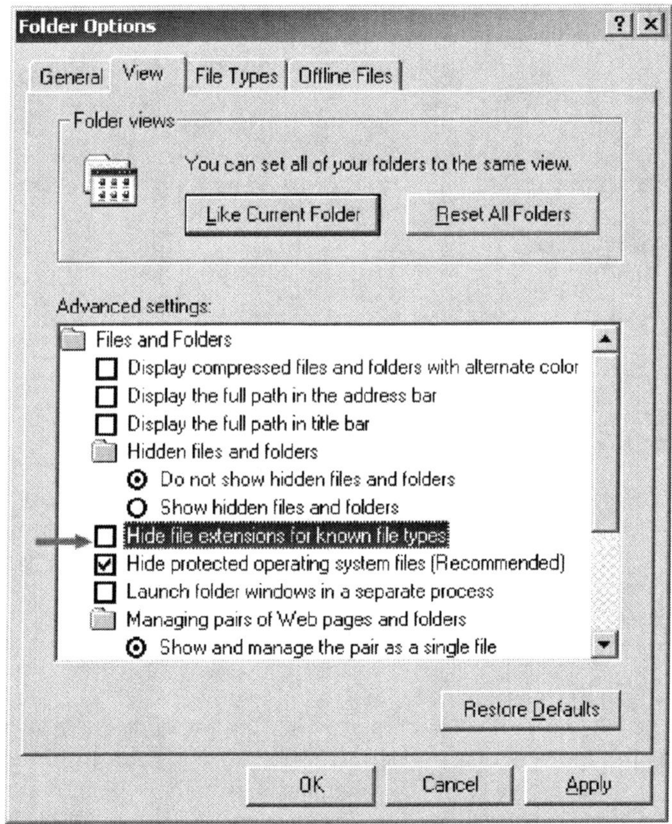

FIGURE 16. Unchecking "Hide file extensions for known file types"

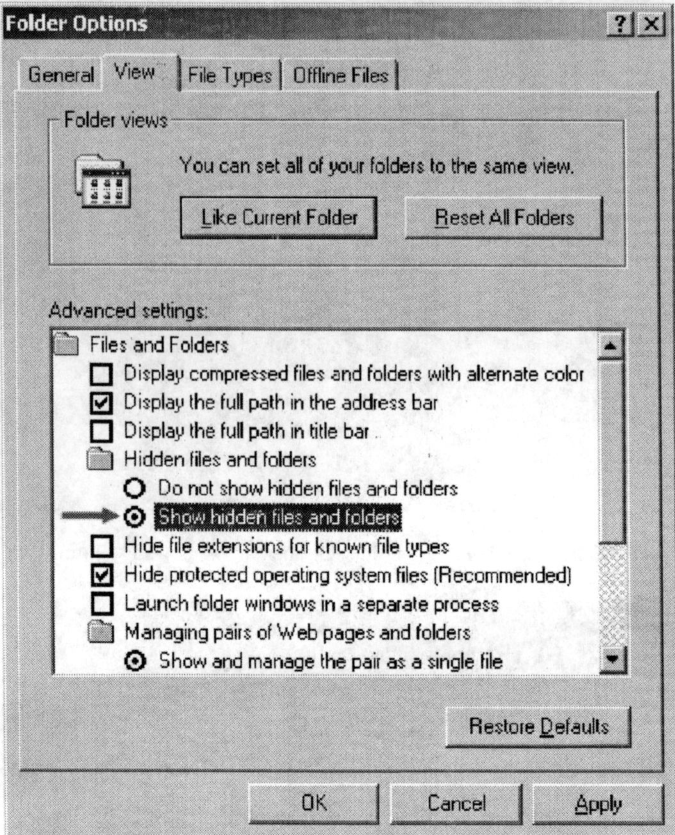

FIGURE 17. Selecting "Show hidden files and folders"

Technical Tip: If you cannot find the Startup folder or folders in the path above this folder, the reason generally is that Windows preferences may be set to hide some folders. To fix this:

- Click *Tools > Folder Options* in Windows Explorer.
- Select the View tab. Select the "Show hidden files or folders" radio button (Figure 17).
- Click *OK*. You should now see the needed folders.

D. Restart EndNote and Word.

- Before you use the EndNote tools in the Word for the first time, close Word, then open EndNote and an EndNote library, then re-open Word and click on one of the EndNote tools. The programs must only be opened in this order the first time you use the tools so that the registry will be updated correctly.

> ⚠ **Alert:** If you are using EndNote 6, make sure that you DO NOT copy the file called "EN6Cwyw.WordXP.wll" because using this file can trigger memory errors in Word XP and Word 2003. If you are using EndNote 7 with Word XP or Word 2003, please note that there is a known issue with the use of EndNote keyboard shortcuts in Word. The EndNote keyboard shortcuts in Word do not function when using the EN7cwyw.dot file that comes with EndNote 7.

What to do if EndNote keyboard shortcuts don't work in Microsoft Word

Sometimes you will notice that even though the EndNote commands in Word are working, keyboard shortcuts such as ALT+7 for the "Find Citation(s)" do not work (refer to the Cheat Sheet at the beginning of the book for more information about keyboard short cuts).

To fix this problem, you need to update the EN7CWYW.DOT file:

- Download the new version of the file from EndNote's web site at http://www.endnote.com/support/faqs/enfaq4.asp.
- Locate the Word Startup folder by following the steps in the previous section. Delete the old EN7CWYW file and paste the new file.

How to fix the Macro Security Warning in Microsoft Word

This is another problem that may prevent you from using the EndNote menu in Microsoft Word. In this scenario, even if the EndNote menu appears under Word, if you try to use any of the commands then you are presented with a macro security warning (Figure 18). This generally happens because your system is trying to prevent Word from accessing EndNote files due to its concern that these files may be harmful to your computer.

To fix this problem, perform the following steps:

- Close all programs running on your computer and open Word again.
- Click *Tools > Macro > Security* in Word (Figure 19).
- In the next dialog box:
  - Select the *Medium* setting in the *Security Level* tab (Figure 20).
  - **Uncheck** "*Trust all installed add-ins and templates*" box in the *Trusted Sources* tab (Figure 21).
  - Click *OK*.

FIGURE 18. Macro security warning in Microsoft Word

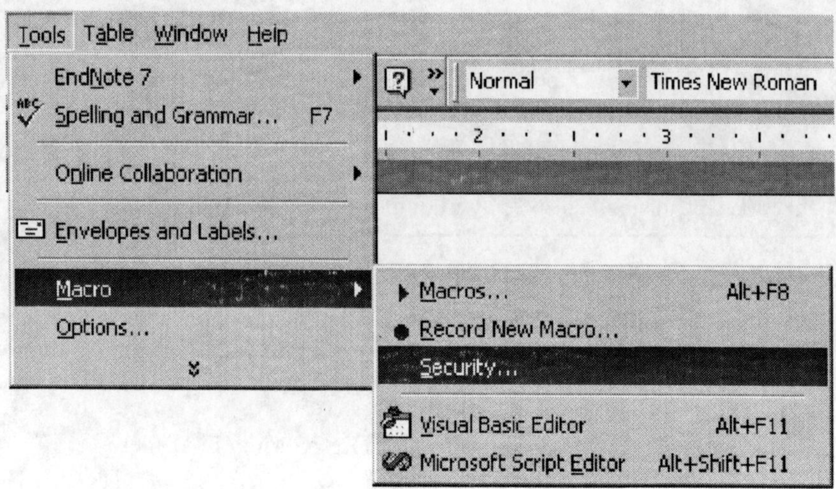

FIGURE 19. Selecting the Macro > Security command

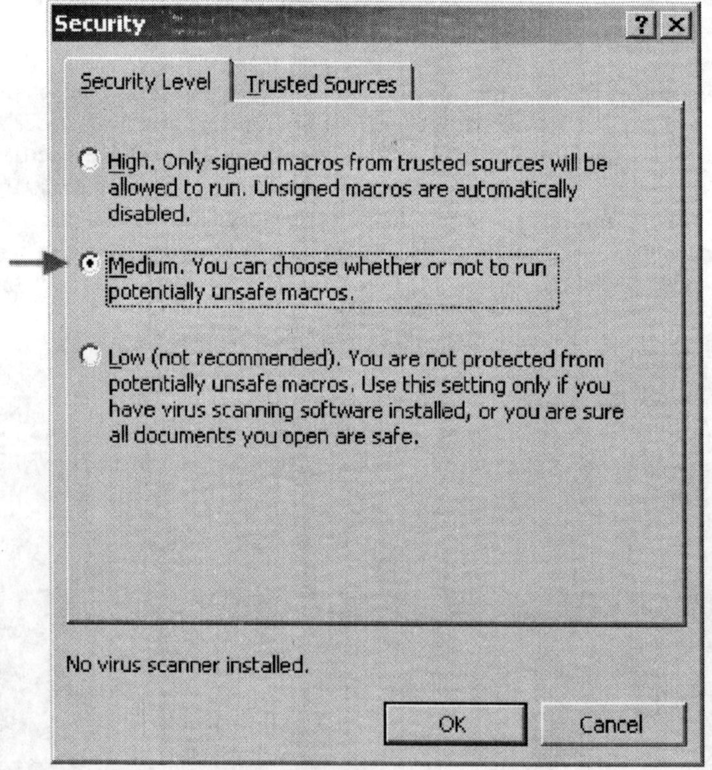

FIGURE 20. Selecting the Medium setting in the Security Level tab

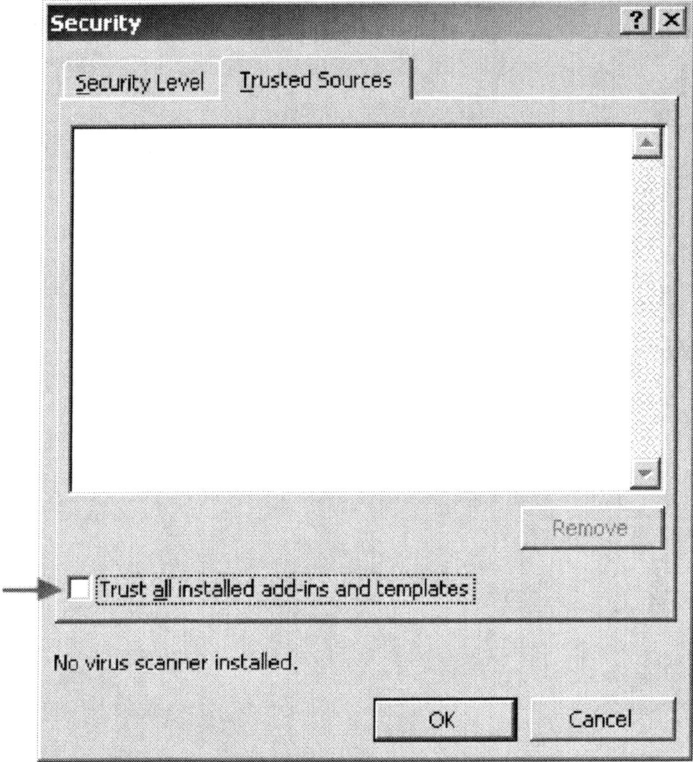

FIGURE 21. Unchecking the Trust All . . . box in the Trusted Sources tab

- Close Word and open Word again. Now you will be presented with a Macro Security dialog box (Figure 22).
- Click *Enable Macros* for each waning box pertaining to EndNote macros. If you are unsure about the safety of a macro, don't install it.
- After you have enabled all the EndNote macros, select *Tools > Macro > Security* again (Figure 19).
- Under the *Trusted Sources* tab, check the box "*Trust all installed add-ins and templates.*" Click *OK* (Figure 21).

This should fix the macro security warning problem, and you should be able to use the EndNote commands in Word now.

Automatically Updating EndNote

The EndNote vendor will at times make additional files available to keep your software updated. These may include feature enhancements or 'fixes' to problems

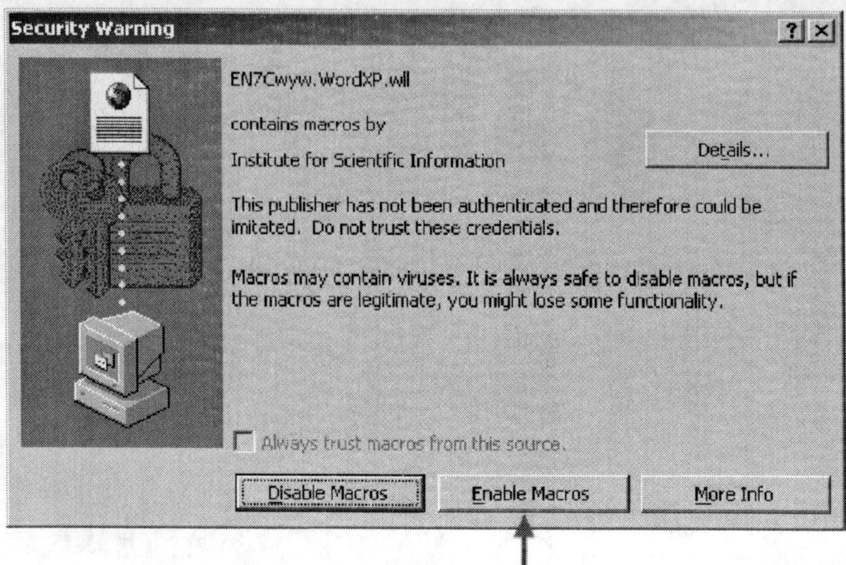

FIGURE 22. The Macro Security dialog box

identified (called 'glitches' in computer lingo). You should periodically update your software to get the latest updates. Remember, you must be connected to the Internet to use this feature.

To update EndNote:

- Start EndNote program;
- Click *Help* > *EndNote updates* (Figure 23).

Follow the instruction on the next screens. The software wizard checks for updates, downloads them and installs them automatically on your computer.

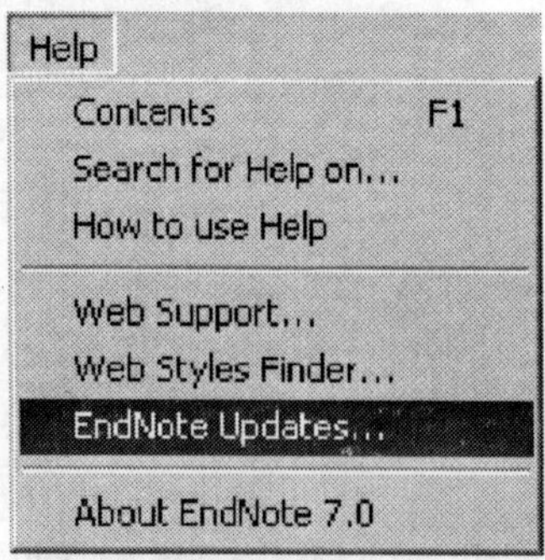

FIGURE 23. Selecting Help > EndNote Updates

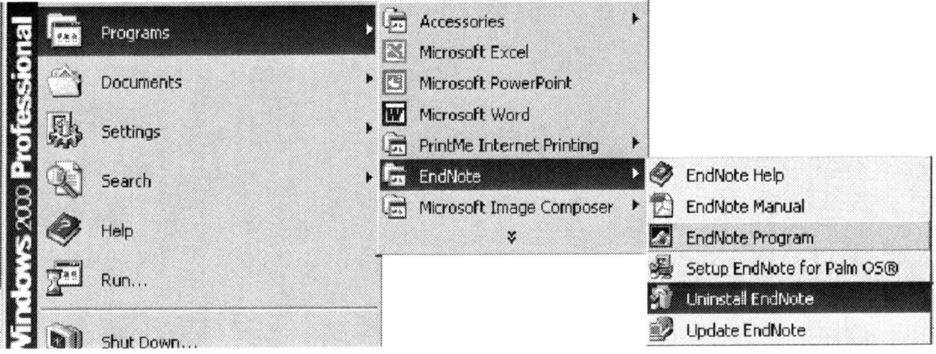

FIGURE 24. Uninstalling EndNote

Uninstalling EndNote

To uninstall EndNote:

- Click Start > Programs > EndNote
- Select *Uninstall EndNote* (Figure 24).

> **Technical Tip:** The uninstall program removes only files, groups, and icons installed by the EndNote installer the last time it was run. For example, It will NOT delete your libraries or any new files you have created. It will also NOT delete folders if they contain files you created.

File Compatibility Issues

EndNote Library

EndNote 7 for windows is fully compatible with all libraries from earlier versions of EndNote and EndNote plus for DOS, Windows, and Macintosh.

EndNote Styles

EndNote 7 can use styles created by EndNote versions 2–6; however, EndNote 7 styles cannot be used by versions prior to version 4. When opening an EndNote style prior to version 4, EndNote 7 opens it as a new, untitled style, which you may save with a new name. The original style remains untouched so that you may still use it with other EndNote versions.

4
EndNote Libraries

Things You will Learn in This Chapter

- The basic features of the EndNote library.
- About working with an EndNote library and performing functions such as creating, opening, closing, saving, and backing-up EndNote library.
- How to sort EndNote library.
- How to customize EndNote library by setting preferences.
- How to recover a damaged library.
- How to merge libraries.
- How to post EndNote libraries on the world wide web.

What is an EndNote Library?

Just as you create files, such as documents in Microsoft Word and spreadsheets in Microsoft Excel, your work with EndNote involves creating files known as EndNote libraries. An EndNote library is essentially an electronic database containing various types of references, for instance, journal articles, books, magazine articles, figures, tables, and so on. Like any other computer file, an EndNote library can be moved, copied, renamed, or deleted using standard Windows commands.

Features of an EndNote Library

The following summarizes the salient features of an EndNote library:

A. Each EndNote reference in a library consists of various **reference fields** such as Author, Title, Year, URL, Publication date, etc. Each field in a reference can accommodate approximately 8 pages of text (32,000 characters), with a total limit of about 16 pages (64,000 characters) per reference.

B. Each library can have a maximum size of 32 MB or 32,267 record numbers in it, whichever comes first. For most personal users, this limit is rarely reached. If you anticipate that you will need to create larger files, you should consider other reference management software such as, ProCite® or Reference Manager® which don't have a file size limit.
C. Each reference added to a library is assigned a unique record number that never changes for that reference in that library. You cannot edit this number.
D. Even if you delete some references, record numbers in the library associated with the references are never used again. So, for example, if you imported 32,000 references in a library and then deleted all of them, still you only have space for only 267 more references in that particular library.
E. If your library includes image files associated with references, EndNote automatically creates a separate folder to store these images. This folder is named "*Library.DATA*" and is found in the same location as the library. If you move or copy the library, you must move or copy this "*.DATA*" folder along with the library otherwise you will lose access to images from your library.

⚠ **Alert:** If you are backup EndNote library onto a floppy disk or another storage medium such as CD or a USB drive, make sure to also back up the "*.DATA*" folder.

If you e-mail a library to someone, and your library contains images or attached files, remember to e-mail the associated "*.DATA*" folder as well. Generally, the easiest way to send via e-mail is to 'zip' the library and the "*.DATA*" folder into a single file using a file compression utility such as WinZip® (http://www.winzip.com) or Stuffit (http://www.stuffit.com).

F. EndNote 7 offers full compatibility between Windows and Macintosh platforms. Libraries created with EndNote 7 for Windows can be read by EndNote 7 for Macintosh and vice-versa without any conversion.

Technical Tip: What happens to the size of the EndNote library when an image or a PDF file is attached to a reference? Does the size of the PDF file or the image attachment count toward the 32 MB size limit of an EndNote library? The answer is: no. Since these files are stored in the separate "*.DATA*" folder, the size of the attached file has no impact on the size of the EndNote library itself.

Working with an EndNote Library

If you already have an EndNote library in your computer, you can use it to follow the examples in this chapter. For your convenience, I have included a sample

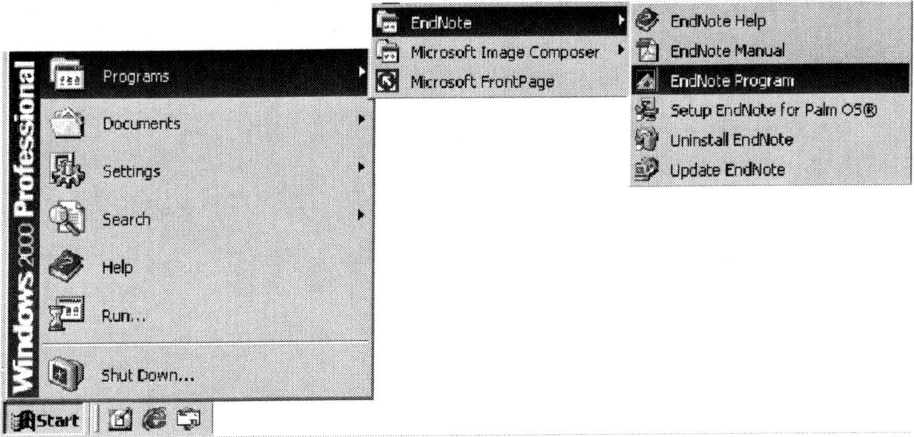

FIGURE 1. Starting EndNote program

library for you in the CD accompanying this book to learn EndNote functions and commands till you create your own.

Opening a Library

- Double-click the library name and it will start the EndNote program and open the library
  OR
- Start EndNote using the *Start > Programs > EndNote > EndNote Program* command (Figure 1).

You will be presented with a dialog box with options (Figure 2):

- If you have already opened a library before, you will see the name of that library in the dialog box. Click *OK* to open this library.
- If you would like to open some other library, click *Browse* to locate the file and then click *OK*.
- You can also create a new library by clicking on the radio button "*Create a new EndNote library.*"

Technical Tip: When trying to open or work with an EndNote library, sometimes you will get an error message—"The file is locked or in a locked volume. No changes will be saved." This generally means that this file is write-protected. This often happens if you 'burned' this file on a CD and then copied it to a computer. To fix this:

- Right-click on the file name. Select *Properties* from the menu (Figure 3).
- In the next screen, make sure the General tab is selected. Uncheck the "Read-only" box (Figure 4). Click *OK*.

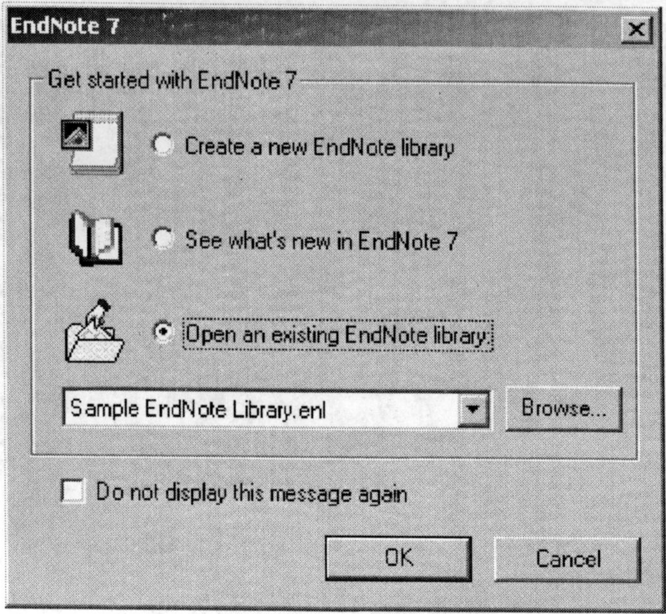

FIGURE 2. Opening an EndNote library

Understanding the EndNote Library Window (Figure 5)

The EndNote library window displays a list of references with each reference in a separate row. Various references fields are displayed in vertical columns. By default, the following columns are displayed for each reference:

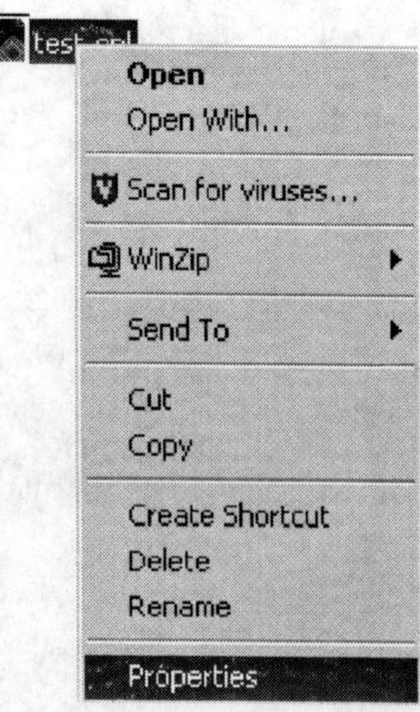

FIGURE 3. Selecting Properties from the right-click menu

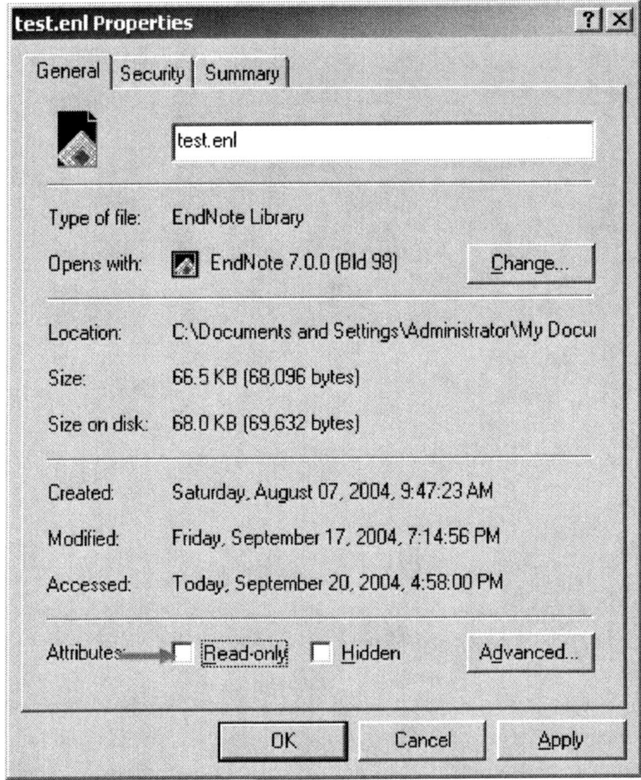

FIGURE 4. Unchecking the "Read-only" box

(a) A paper clip indicating a file or image attached to a reference;
(b) Author, indicating the first author's last name;
(c) Year;
(d) Title;
(e) URL.

You can modify the columns displayed and their sequence according to your preference, as you will learn later in this chapter. Figure 5 illustrates other important features of the EndNote library window.

Creating a New Library

To create a new library:

- Click *File* > *New* in the EndNote window. EndNote opens a dialog box (Figure 6).
- Select a location where you want to save your library. Enter a name for your library. You do not need to include the extension ".enl" in the name—EndNote automatically does it for you.
- Click *Save* to save your library.

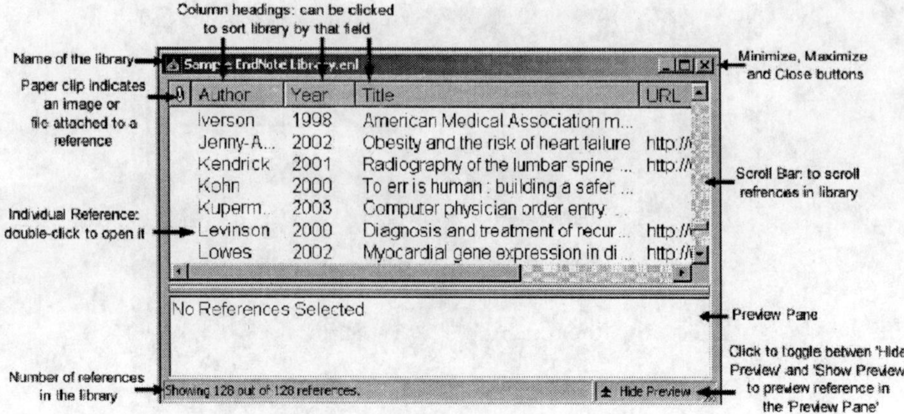

FIGURE 5. EndNote library window

Sorting the EndNote Library

By default, references are sorted in ascending order by the last name of the first author. It is very easy to sort the library by any other reference field (such as by the year or alphabetically by the title) simply by clicking on the name of that column (Figure 5).

For example, to sort by the year, click on the "Year" column and your references are sorted in an ascending order by the year. If you click on the column name again, your references are now sorted by the year in a descending order.

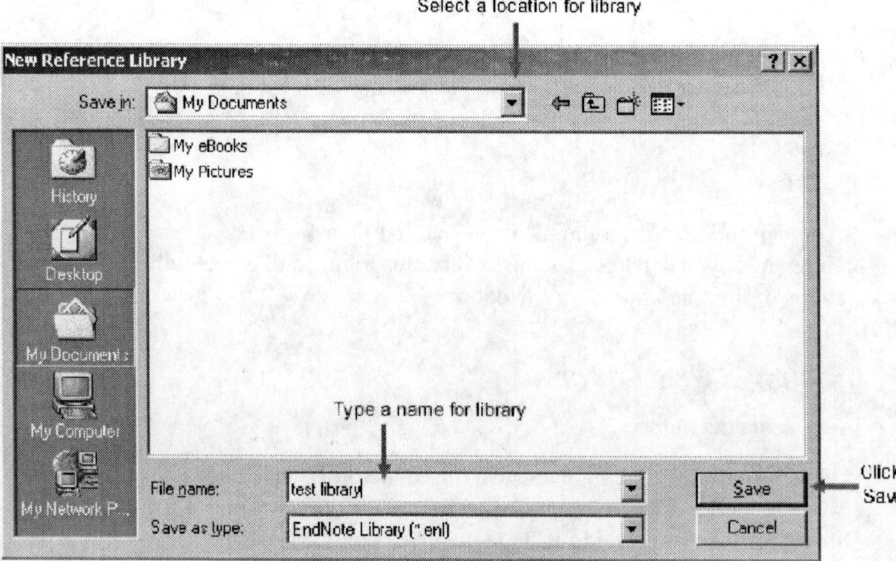

FIGURE 6. Creating a new EndNote library

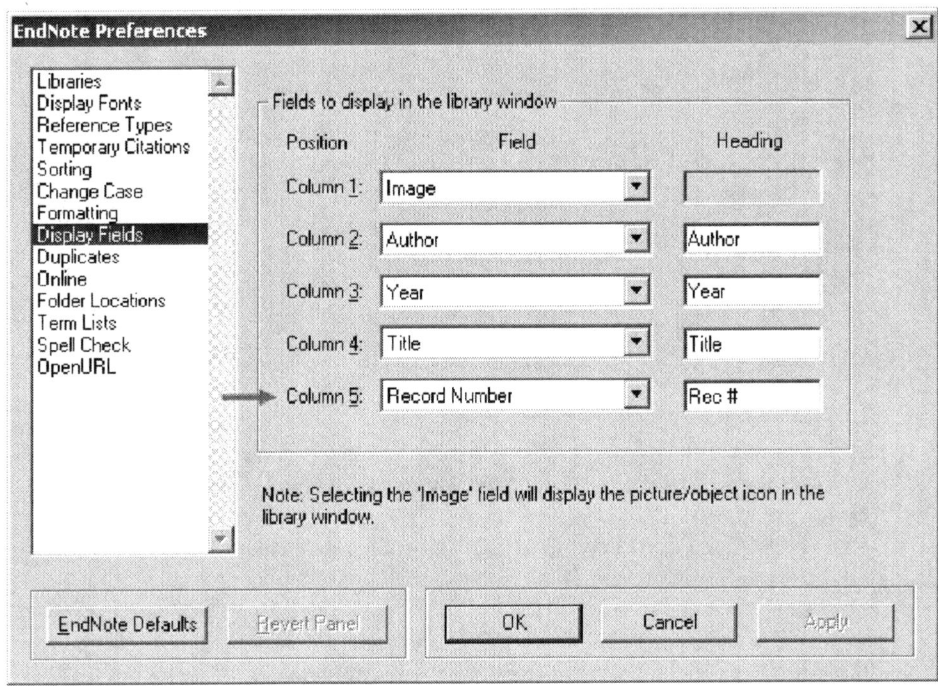

FIGURE 7. Displaying record number in an EndNote library

> **Technical Tip:** There is no direct way in EndNote to sort a library by the date on which you created records. If you would like to sort the library by this date, one way of accomplishing this is to sort the library by 'record number' as the most recently created references should have the highest record number. To do this:
>
> - Add "record number" as one of the display fields in the library.
>   - Click *Edit > Preferences* in EndNote.
>   - In the next screen, select Display Fields. Click on the dropdown menu in one of the columns and select "Record Number" from the list (Figure 7).
> - Your library should now display record number as one of the fields. Click on this field to sort records in ascending or descending order (Figure 8).

Navigating the EndNote Library

To browse through references in an EndNote library, you can choose any of these options:

- Scroll through the list of references in the library by clicking on the vertical scroll bar.

Author	Year	Title	Rec #
	2003	Thanks for no memory	#10
	2003	Junk DNA: An ancient he...	#11
Kohn	2000	To err is human : building...	#12
United ...	2001	Making health care safer...	#13
Austin	2003	Information systems for h...	#14
Health ...	2001	Back to basics : foundati...	#15
Petrovi...		Neuro-fuzzy modelling in ...	#16
Rosen...	1993	The Design and Impleme...	#17

FIGURE 8. EndNote library sorted by the record number field

- Use the ↑ or ↓ symbols on the keyboard to move up or down the reference list.
- **Quick keys:**
  - The HOME and END keys go to the first or last reference respectively.
  - The PAGE UP and PAGE DOWN keys move up or down one screen of references.
- Typing a letter selects the first matching reference. The matching depends upon the sort order of the library. For example, if your library is sorted by author field, typing selects the first reference matching the author's last name.

Previewing a Reference

You can easily preview details about a reference in a preview pane by simply clicking on it to highlight it. Click the Show Preview/Hide Preview button at the lower right corner to preview references in the preview pane (Figure 9).

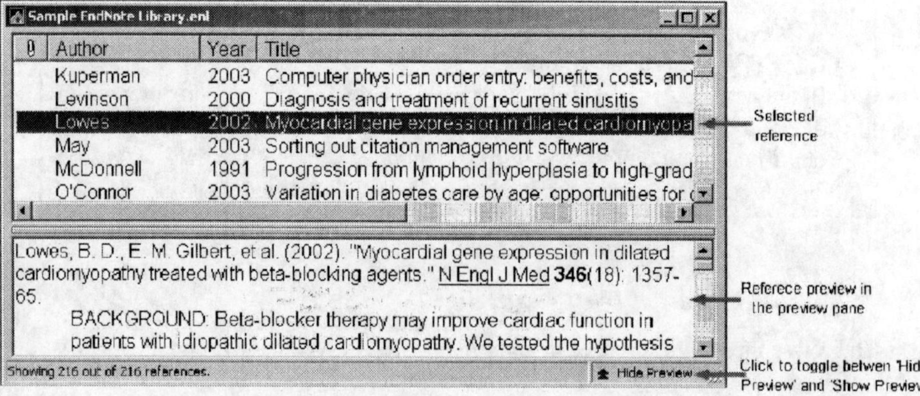

FIGURE 9. Previewing a reference

Only one reference is displayed in the preview pane at a time.

To preview multiple references:

- Select multiple references by clicking and highlighting. (Tip: hold CTRL key while clicking to highlight multiple references.)
- Click *Edit > Copy Formatted* to copy the formatted references to the Windows clipboard.
- Create a new document in word processor. *Paste* the formatted reference into the document to preview.

Technical Tip: It may be useful to configure the library window such that the main window displays the bibliographic data about the reference (such as author, title, date etc.) and the preview pane displays the abstract. This would allow you to quickly browse through your references just by clicking on them without having to open individual references. Perform the following steps to create this configuration:

- Click *Edit > Output Styles > New Style* (Figure 10).
- In the next window, click *Templates* under *Bibliography* (Figure 11).
- Click on the "Generic" template.
- Click on *Insert Field* dropdown menu.
- Select *'Abstract'* from the menu.
- Click on *File > Save as* and give this style a name (such as 'abstract').

You should now see abstracts in the preview pane. As long as this is the selected output style in EndNote, you will see abstracts in the preview pane. See Chapter 8 for a detailed discussion of output styles.

Copying Between Libraries

If you want to copy some references from one library to another:

- Open the libraries you want to copy to and from.
- Highlight the references you would like to copy (using CTRL to select multiple references).
- Click on any part of the selection and use the mouse to drag the selection to another library or use the *Copy* and *Paste* commands to transfer references.

Setting Library Preferences

EndNote allows you to customize an EndNote library by setting EndNote preferences. Some useful customization features include:

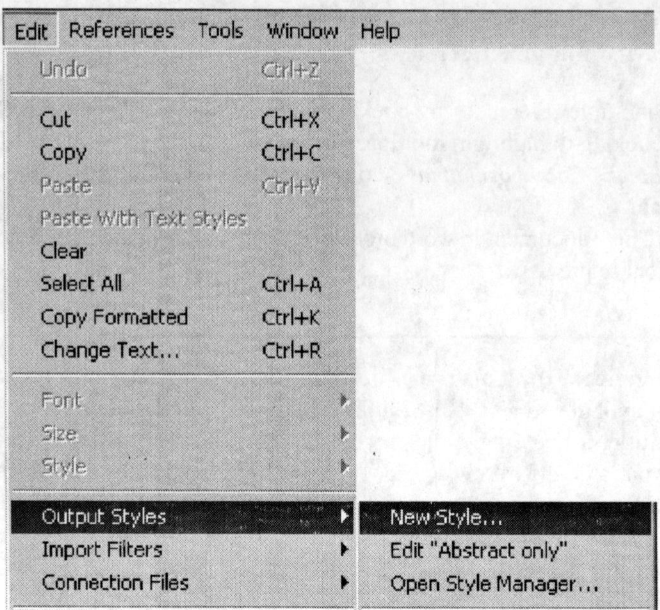

FIGURE 10. Selecting output style

Setting Fonts in the Library

There are two types of font settings in the library:

(A) The Library Display Font—used by the list of references in the library's main window;

(B) The General Display Font—used by the preview pane.

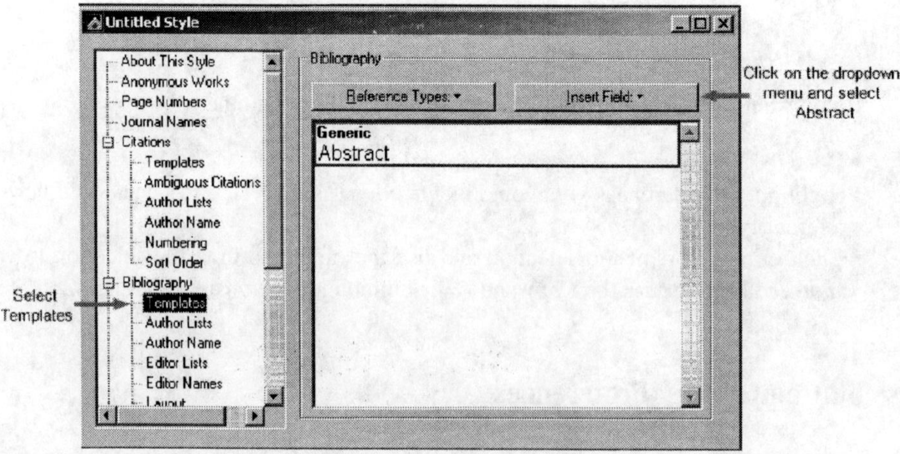

FIGURE 11. Creating an output style to display an abstract in the preview pane

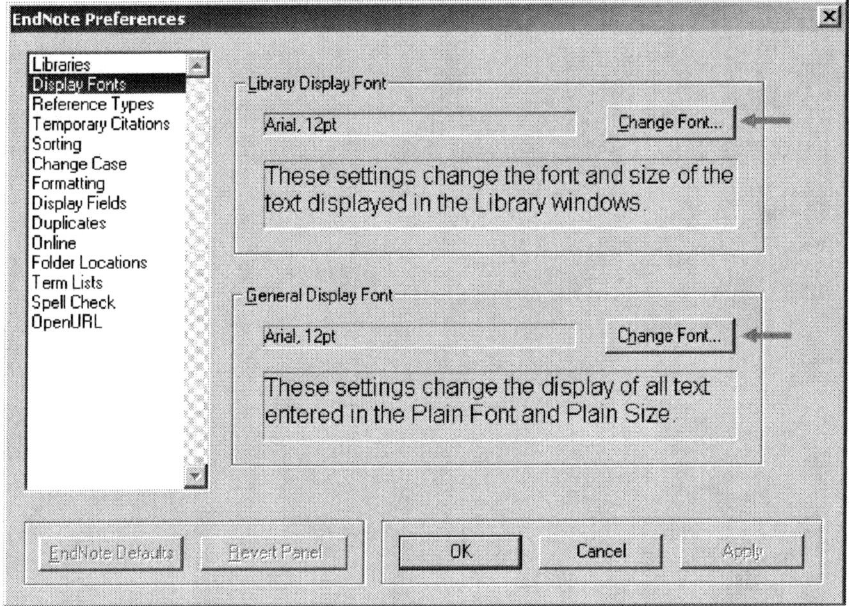

FIGURE 12. Setting EndNote library fonts

⚠ **Alert:** Changing these library display fonts does not change the font in the bibliography created using the EndNote library or the font of your paper.

To change the display font:

- Click *Edit > Preferences*.
- Click the *Display Fonts* option (Figure 12).
- Click the *Change Font* button in the "Library Display Font" (top section) and the "General Display Font" at the bottom section of the screen. Select a font type and size in the next window.
- Click *OK* to save the changes and close the Preferences dialog box.

Setting a Default Library

You can specify the library you want to open automatically when EndNote starts. This library is called default library.

To assign a default library (Figure 13):

- Open the library or libraries that you would like to set as default library. **You must have a library open to set it as a default library.**
- Click *Edit > Preferences*.
- Click the *Libraries* option.

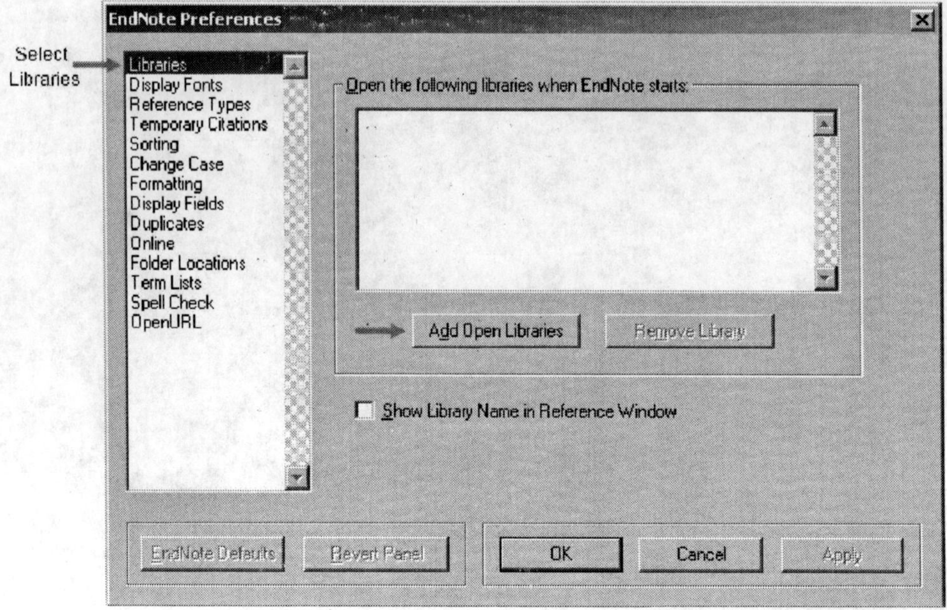

FIGURE 13. Setting a default library

- Click *Add Open Libraries* button and all of the currently open libraries will be added to the list of default libraries.
- Click *OK* to save changes.

Similarly, you can click the *Remove Library* button to remove a library from the list of default libraries.

Setting the Fields to Be Displayed in the Library

By default, the following five fields are displayed in sequence from left to right in the main library window: a paper clip if an image or a file is attached to the reference, Author, Year, Title, and URL. You can customize the look of the library by changing the order of the fields displayed, adding other fields, or changing the names used for the column headings in the library window.

To customize the display fields (Figure 14):

- Click *Edit > Preferences*.
- Click the *Display Fields* option.
- Choose the desired fields from the Field list for positions 1 through 5.
- If you would like to use fewer than 5 display fields, select Unused for those positions. For example, in Figure 14, I have designated column 4 as 'unused.'

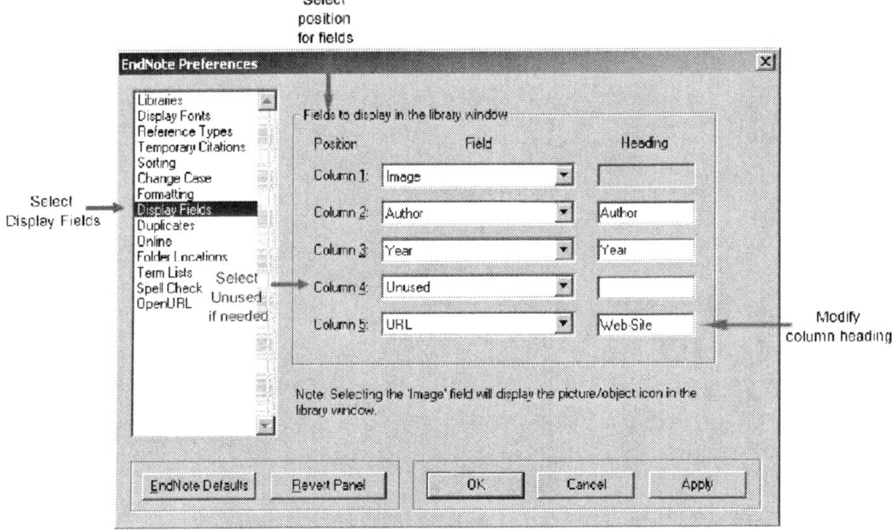

FIGURE 14. Setting display fields in an EndNote library

- If you would like to change the name of a field for display in library, change the name under the heading. For example, in Figure 14, I have changed the name of the column 'URL' to 'website'.

Recovering a Damaged Library

Occasionally, you will have a damaged library giving you an error message when you try to open it or work with it. EndNote provides a 'recover library' utility to recover damaged libraries. Just remember that you may not be able to recover the data completely and the best way to protect yourself from the lost data is to create regular backups of your files on a CD or any other storage media.

To recover a damaged library:

- Close any open libraries (but leave the EndNote program running).
- Click *Tools* > *Recover Library* from the menu (Figure 15).
- Read the information in the next dialog box and click *OK* (Figure 16).
- In the next screen, locate and select the library that needs to be recovered and click *Open*.

EndNote creates a copy of the library in the same folder as the original library. This new copy has the suffix "-Saved" added to its name. For example, if you were recovering the "test" library, it will be recovered as "test-Saved."

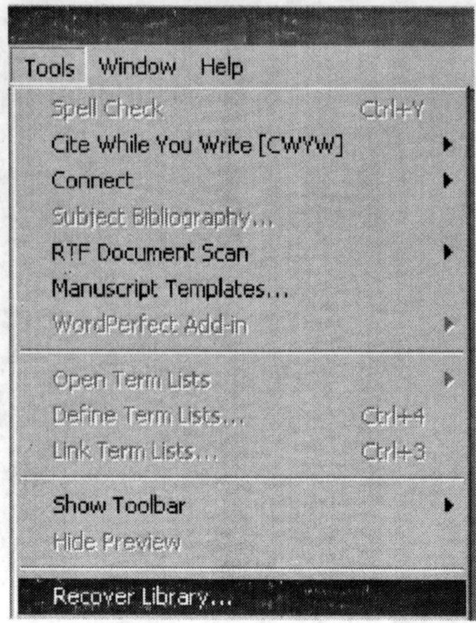

FIGURE 15. Recovering a damaged library

Alert: The *Recover Library* command does not create a new "*.DATA*" folder. So in the above example, if your library contains attached images or files, you will have the recovered library saved under the name "test-Saved" whereas the "*.DATA*" folder is still named "*test.DATA*".

For the EndNote library to be able to access images, the library and the "*.DATA*" folder must have the same name.

Therefore, you should either rename the recovered library to the original library name ("test") or rename the "*.DATA*" folder to match the new library's name ("test-Saved.DATA).

Helpful Hints About a Recovered Library

- Often the size of the 'recovered' library is smaller than the original. This is normal and you do not need to be alarmed.

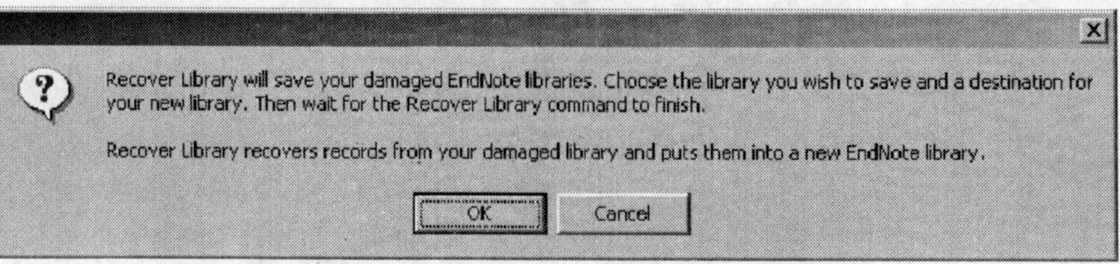

FIGURE 16. The Recover Library dialog box

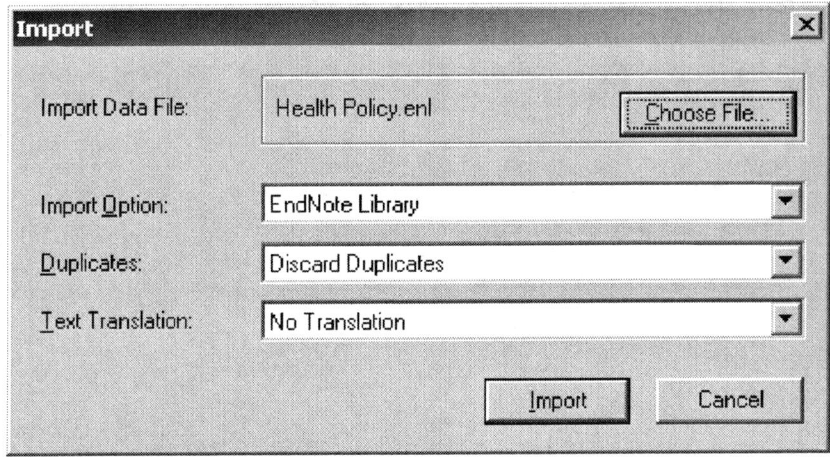

FIGURE 17. Merging libraries using the Import mothod

- The recovered library may also contain references that were recently deleted from the original library. If so, you may need to delete them again from the recovered library.
- Term lists (see Chapter 5 for more about term lists) are not recovered with the library and you will need to rebuild them.

Merging EndNote Libraries

There are two ways to merge EndNote libraries:

(A) Import one library into another using the *Import* command. To do this:
  - Click *File > Import* EndNote.
  - Select appropriate choices in the next screen (Figure 17).
    - *Import Option* should be EndNote library.
    - Choose desired *Duplicates* setting.
    - Click *Choose File* to select an EndNote library you would like to merge.
  - Click *Import*.

(B) Copy references from one library into another. To do this:
  - Select a reference by clicking on it, or select multiple references by holding down the CTRL key while selecting.
  - Use *Copy > Paste* command or drag and drop to bring references from one library to another.

Technical Tip: Generally it is better to use the *Import* command method especially if you will be importing many references because the import method automatically filters out duplicate references during the process.

⚠ **Alert:** When you add references to a library by using either method, the newly added references are assigned new record numbers. For example, a reference that was #12 in the old library may be assigned #225 in the new library.

If you have any documents that have citations linked to #12 in the old library, you won't be able to find this citation in the new library.

If you have any documents using references from either library, you must make a backup on the library before performing a merge.

Publishing an EndNote Library on the Web

You may want to post your EndNote library on the Web in a format that is searchable and allows you to regularly update the library database on the Web. This is especially helpful if you have a large collection of references that you would like to share with other people.

EndNote's website mentions that EndNote 7 supports web publishing. Essentially, this consists of 'exporting' a library as an HTML or XML page but the result is quite limited in functionality. Using the *File > Export* command, EndNote will export selected references from a library into HTML or XML only in a selected output style. For example, if you selected 'Life Sciences' as your output style, then only the reference fields pertinent to the life sciences output style will be exported, not the entire database. In addition, this HTML or XML file cannot be searched.

There used to be a product called Reference Web Poster provided by EndNote vendor to publish EndNote libraries on the web, but this product is not distributed or supported anymore.

If you are seriously interested in posting your EndNote library on the Web, you will need to consider options other than EndNote's *Export* command. The most important consideration in selecting an option is the level of technical expertise you have and the amount of extra money you are willing to spend.

One common and relatively easy option is to export your EndNote library to another reference management software that directly supports posting reference libraries on the web. The advantage of using this method is that it is very simple and does not require much specialized technical skill in dealing with databases or configuring servers. The disadvantage is that you may need to spend extra money to buy a second software application. Some of the software programs supporting posting reference libraries on the Internet include RefWorks, BiblioWeb—a product of BiblioScape—and Reference Manager.

Using RefWorks (http://www.refworks.com) is an easy and cheap option. RefWorks is a **web-based** reference database; therefore, the libraries you export from EndNote to RefWorks are automatically available on the web.

RefWorks requires a paid-subscription membership to publish the library on the web. It then allows you to give read-only access to whomever you want with a special password. People with read-only access can do extensive searches of your

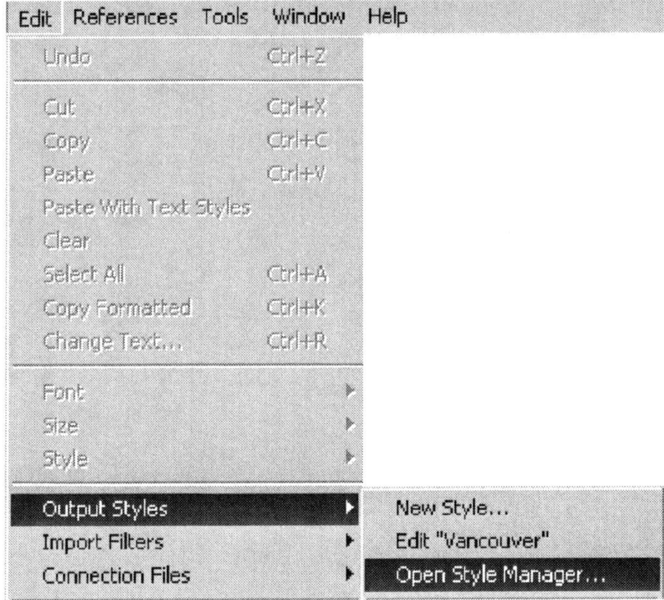

FIGURE 18. Opening Style Manager

library and output references into virtually any format. RefWorks offers a free 30-day trial membership.

Publishing EndNote libraries on the web using RefWorks involves two steps:

A. Export EndNote records to a text file using a specific output style.

- First you should make sure that the **RefMan (RIS) Export** is selected as the output style in EndNote. To do this, click on *Edit > Output Styles > Open Style Manager* (Figure 18).
- In the style manager, select RefMan (RIS) Export style (Figure 19). Close the style manager window.
- In the Output Style menu, make sure the RefMan (RIS) Export style is checked by clicking on it (Figure 20).
- Select the references you want to export by clicking them (hold CTRL key while clicking to select multiple references). Under the *Reference* menu select *Show Selected* (or *Show All* if you wish to export the entire database).
- Click *File > Export* in EndNote.
- A dialog box appears for you to select the location where you want to save the file. Type a file name and select a location for importing into RefWorks. Select Text as your Save As type and click *Save*.

B. Import the text file into RefWorks

- Go to RefWorks website and log in using your user name and password.
- Under the References menu, click *Import* (Figure 21).

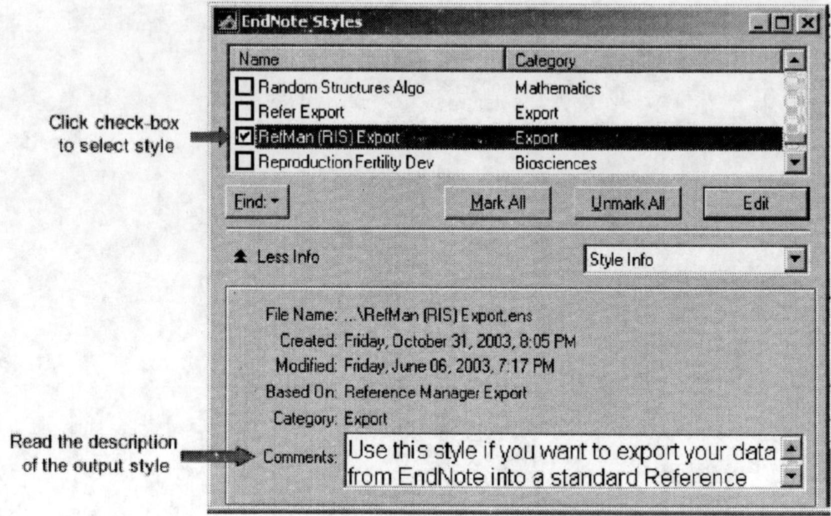

FIGURE 19. Selecting RefMan Style

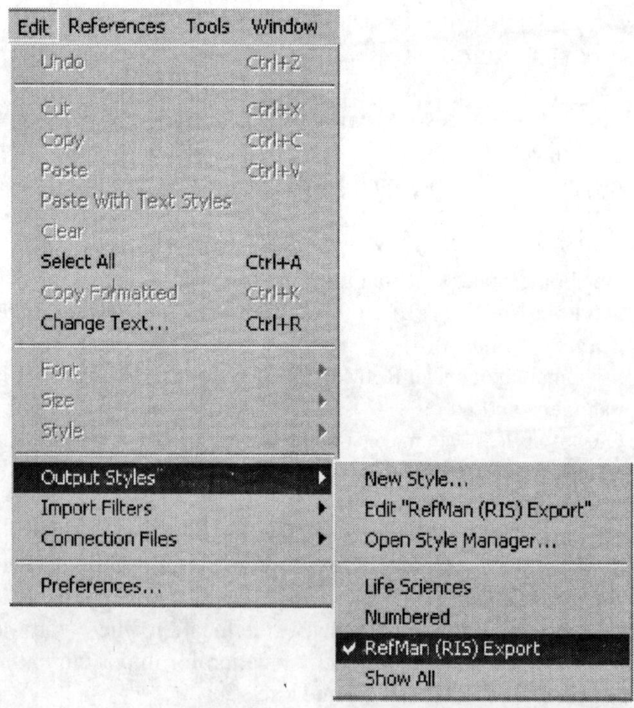

FIGURE 20. Selecting RefMan (RIS) Export style in the Output Styles menu

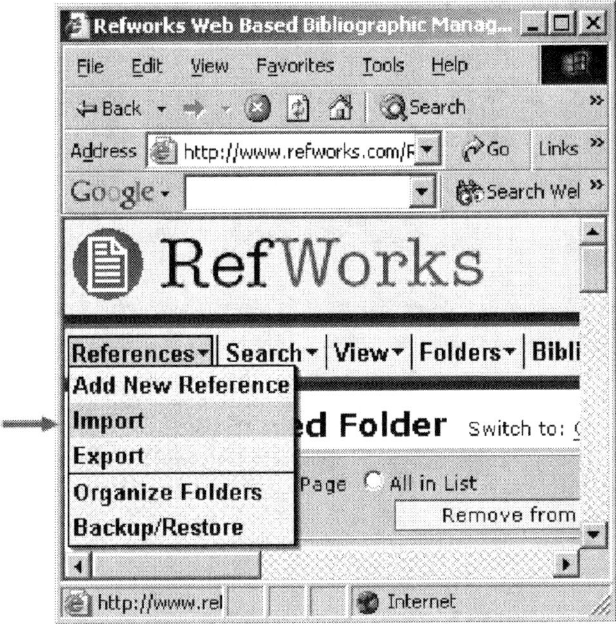

FIGURE 21. Selecting Import from the References menu in RefWorks

- Select *Desktop Biblio. Mgt. Software* as your *Import Filter* and *EndNote* as your *Database*. Browse for and select the file you just created from EndNote (Figure 22). Click *Import*.
- Your selected references are now available online on the website of RefWorks (Figure 23).

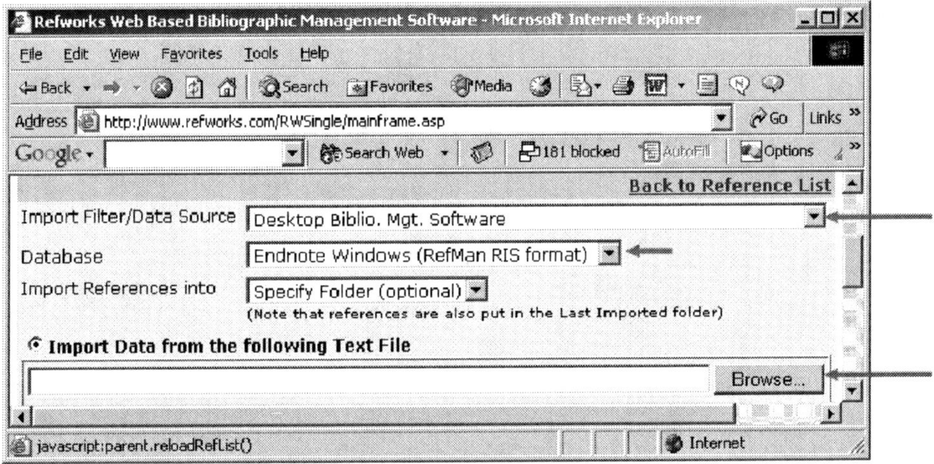

FIGURE 22. Selecting Import options in RefWorks

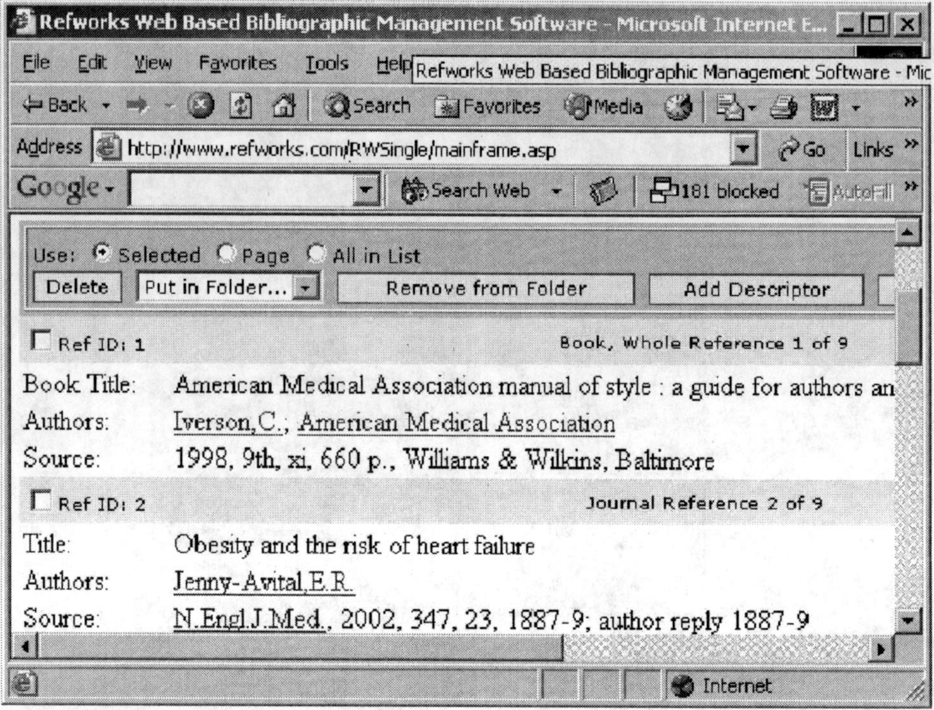

FIGURE 23. RefWorks website showing imported EndNote references

> **Technical Tip:** When importing large amounts of references, RefWorks recommends that you import no more than 2500 records (or 3MB of files) at a single time. This will ensure a successful import.

5
Entering References into EndNote Library

Things You will Learn in This Chapter

- An overview of different methods of entering references in EndNote with manual entry typing as the focus of this chapter.
- About reference fields and reference types.
- How to customize reference types.
- How to create a new reference and set a default reference type.
- Appropriate methods and tips for entering data in various reference fields to get accurate results in the bibliography.
- About term lists, their advantages, and how to use them to facilitate data entry.
- How to enter Figure and Chart and Table reference type.
- How to enter special characters such as â or é in EndNote references.
- How to perform spell-check in EndNote.
- How to download references from the websites of various journals.
- How to import references from other reference management programs into EndNote.

Introduction

In Chapters 3 and 4, you learned to create and manage EndNote libraries. This chapter will demonstrate how to enter references in a library. There are a variety of methods of entering new references into an EndNote library:

1. Manual entry by typing.
2. Downloading references from websites of various journals, such as the New England Journal of Medicine, and Science, directly into EndNote.
3. Importing references from other reference management programs such as ProCite and Reference Manager.
4. Importing references from Internet databases.

This chapter focuses on the first three methods. Chapter 7 discusses in detail the methods to import references from Internet databases.

Technical Tip: My preference is to use manual entry by typing only when I cannot download a reference directly from an Internet database (such as PubMed for a journal article or the Library of Congress for a book) or from a journal website. The main advantages of downloading from online sources are: not having to type, avoiding typographical errors during the entry, ensuring that the correct reference information (e.g. author and journal name) is downloaded, and importing extra information about the reference, such as keywords, URL, abstract, and so on.

Reference Fields and Reference Types

Each EndNote reference consists of various **reference fields** containing the information about a reference. Some reference fields for example, Author, Title, Year, URL, Publication Date etc., are pertinent to citing of a reference in a bibliography. Other fields contain additional information about a reference such as Keywords (to facilitate searching references in a library), and URL (to make it easy to find more information about the reference online). You may enter data in as many or as few fields for a given reference.

EndNote has twenty-five predefined **reference types** such as Journal Article, Book, Book Section, Figure, Electronic Source, and so on. A reference type determines the associated reference fields for a reference. Some fields are common across different reference types, for instance, the Author Name field is available in the reference type Journal Article as well as Book. Other fields are unique among reference types, for instance, the Journal Article reference type has the Journal Name field, and the Book reference type has the field Book Publisher (Figure 1). EndNote also provides one Generic and three additional "Unused" types so you can define your own reference type.

Table 1 lists various reference types available in EndNote.

Technical Tip: Each field in a reference can accommodate 8 pages of text (32,000 characters), with a total limit of about 16 pages (64,000 characters) per reference.

Customizing Reference Types

Reference fields for each reference type can be modified or deleted. You may also add new fields to a reference type—up to a maximum of 40 fields per reference

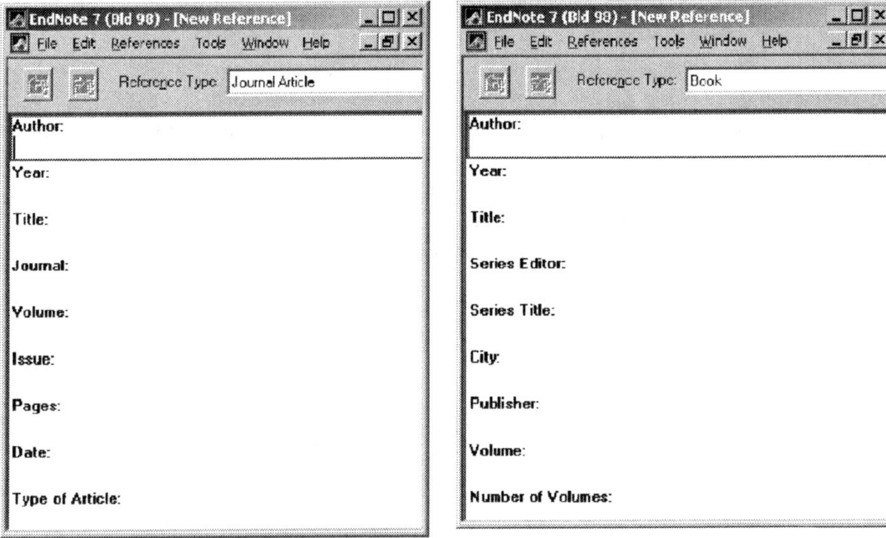

FIGURE 1. Reference fields in different reference types

type. All but the generic reference type can be modified to add, delete, or rename fields according to your preference.

To modify reference types

- Click *Edit* > *Preferences*.
- In the next screen, click to highlight the *Reference Type* heading. Click the Modify Reference Type button (Figure 2).
- The next screen displays the Reference Types table. Edit various fields in the table as necessary (Figure 3).
  - **To rename a field:** Within the column for that reference type, find the name that you want to change, click on it, and type a new name.

TABLE 1. Various reference types available in EndNote

1 Generic	10 Thesis	19 Patent
2 Journal article	11 Report	20 Hearing
3 Book	12 Personal communication	21 Bill
4 Book section	13 Computer program	22 Statute
5 Manuscript	14 Electronic source	23 Case
6 Edited book	15 Audiovisual material	24 Figure
7 Magazine article	16 Film or Broadcast	25 Chart or Table
8 Newspaper article	17 Artwork	26 Equation
9 Conference proceedings	18 Map	27 UNUSED 1/2/3

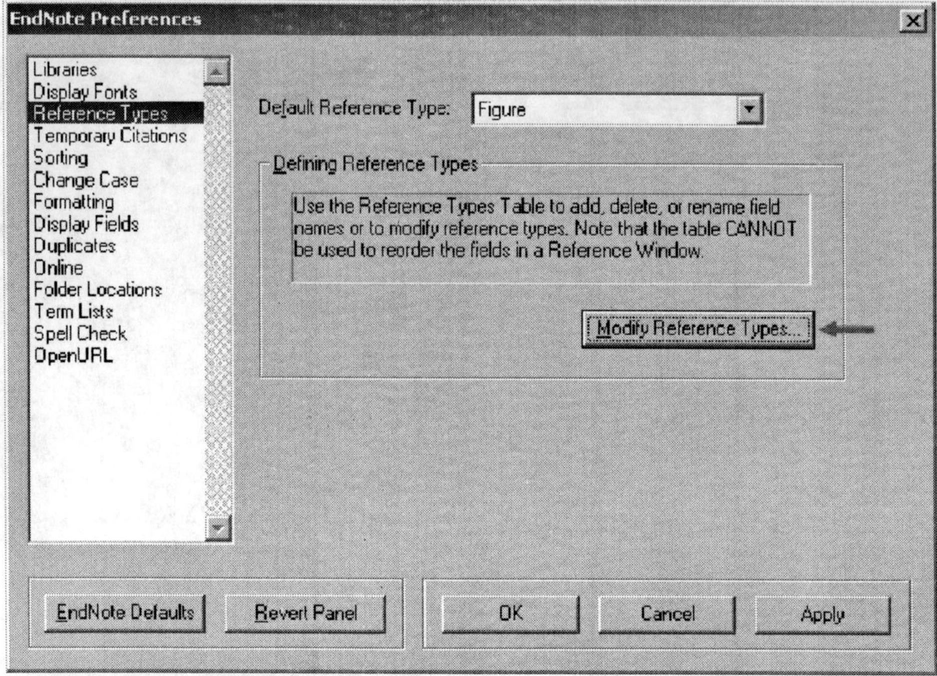

FIGURE 2. Customizing reference types

Edit Reference Types

Generic	Journal Article	Book	Book Section
Author	Author	Author	Author
Year	Year	Year	Year
Title	Title	Title	Title
Secondary Author		Series Editor	Editor
Secondary Title	Journal	Series Title	Book Title
Place Published		City	City
Publisher		Publisher	Publisher
Volume	Volume	Volume	Volume
Number of Volumes		Number of Volumes	Number of Volumes
Number	Issue		
Pages	Pages	Number of Pages	Pages

OK Cancel

FIGURE 3. Screen to edit reference types

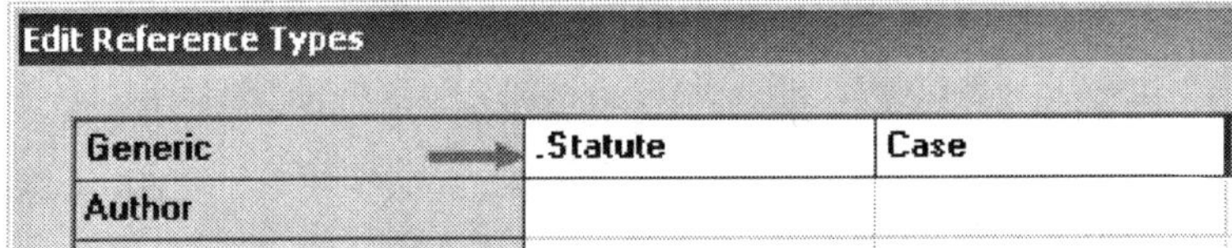

FIGURE 4. Hiding Statute reference type

- **To add a field to a reference type:** Look at the field names in the Generic column and find the one with the most similar meaning. Find the column for the reference type you want to modify. Click in a blank cell and type the name for the field.
- **To delete a field from a reference type:** Simply delete the name of the field from the reference type. If you delete a field that has information in it, the information is not lost. The field is displayed under the generic name.
- **To add a new reference type:** Scroll across to the far right of the table and select one of the "Unused" column headings. Type a name for the new reference type. Add new fields as necessary.
- **Hiding reference type:** You may wish to hide a reference type so that it no longer appears on the dropdown list in the reference window. To do so, simply add a **period (.)** in front of the reference type button and it will no longer display. Figure 4 demonstrates hiding the Statute reference type.

• Click *OK* to save changes.

> ⚠ **Alert:** The changes made to the reference types table apply to all libraries open on that computer. A limitation of EndNote is that these changes are stored in the Windows system registry. Therefore, if you move your library to a different computer, the modified reference types are not available to you and your references will follow the layout of the reference table of that computer.

Creating a New Reference

To create a new reference in EndNote library:

- Click *Reference > New Reference* (Figure 5) in EndNote.
- This opens an empty reference window for entering reference data (Figure 6).
- Select Reference Type from the dropdown list.
- Enter bibliographic data in various reference fields (see next section for details).
- When you are done, simply close reference to save data and add this reference to your library.

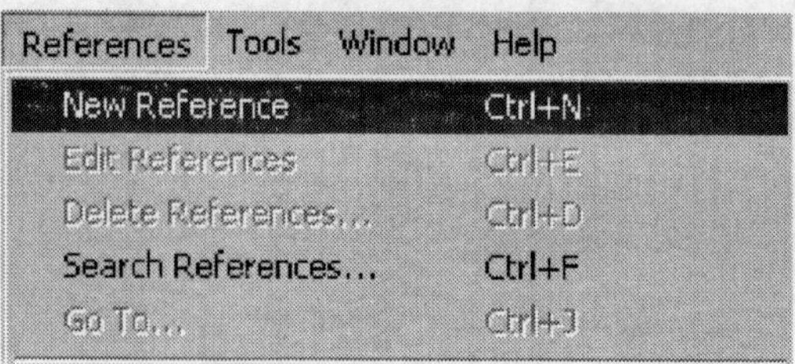

FIGURE 5. Clicking Reference > New Reference

Choosing the Right Reference Type

Selecting the correct reference type is important as it will determine how your reference will be formatted in the manuscript. In addition, the reference type determines which fields appear in the reference window for data entry. For example, a Journal Article reference would have the fields for Journal, Volume, and Issue, whereas a Book reference would have fields for Editor, and Publisher.

Usually you should select a reference type before entering reference data. However, you can change the reference type any time later. The information you already entered in reference is retained and transferred to the corresponding fields of the new reference types.

Setting the Default Reference Type

The default reference type for EndNote is Journal Article; this means when you create a new reference, Journal Article is selected by default in the reference type dropdown list. However, you can set any type as the default so that when you create a new reference, that reference type appears as the default. For example, for

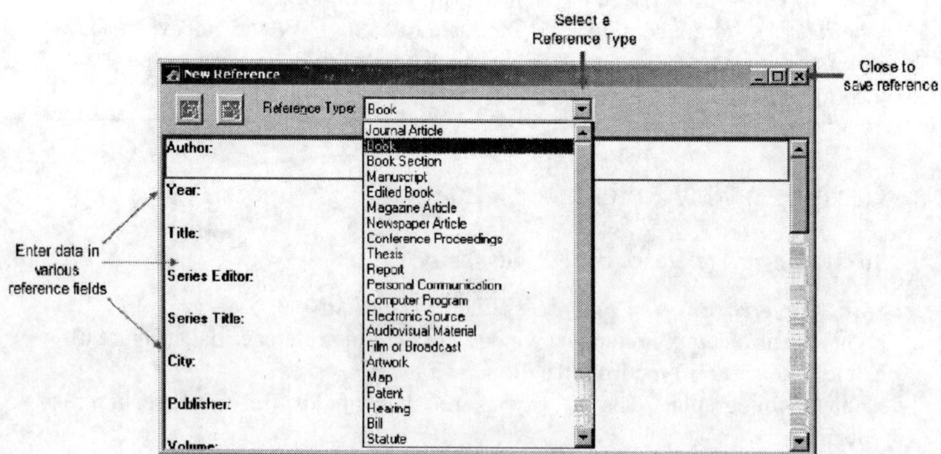

FIGURE 6. Creating a new reference

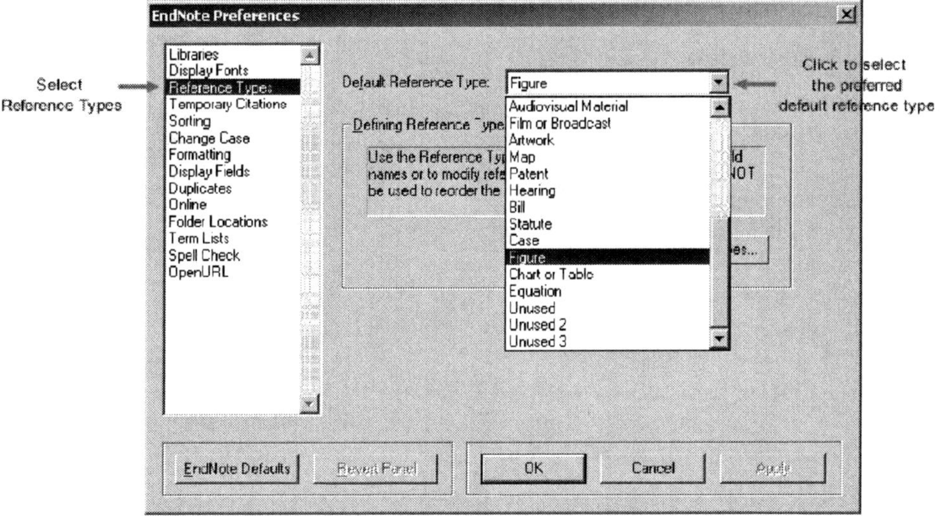

FIGURE 7. Setting the default reference type

writing this book I used EndNote to manage all the images by creating "Figure" type references. Therefore most of the time I was creating a new Figure type reference as opposed to the usual Journal Article type reference. In such a situation, it is very helpful to set the default type to Figure.

Perform the following steps to set a new default reference type:

- Click *Edit > Preferences*.
- Select the Reference Types option (Figure 7).
- Click on the pull-down menu to select the default reference type.
- Click *OK* to save your preference.

Entering Reference Data Manually

Let EndNote Do the Formatting

The appearance of a reference in bibliography is determined primarily by which output style you choose (depending upon the publication you wish to submit your manuscript to), and not merely by the way you enter data in a reference. A general rule is that you should enter only the raw data and let EndNote do the bibliography formatting according to the specification set by the publisher of the manuscript.

Special formatting and punctuation should not be included when you enter references into EndNote. For example, do not include the abbreviation "Vol." along with volume numbers or italicize the journal names.

Guidelines for Entering Data in Various Reference Fields

Author and Editor Names

This is one of the tricky fields to enter data. If you don't enter the data correctly, author names in a bibliography may be formatted incorrectly.

- Enter <u>*one name per line*</u> only.
- Author names can be entered as:
  - Last name, first name OR
  - First name Last Name (no comma).
- If you have a middle name, enter it. Do not abbreviate any names; EndNote can do it for you.
- Enter all author names for a reference. EndNote will truncate the list of authors with "et al." or "and others" depending upon the requirements of the bibliographic style.
- If you do not know all the names, then the last author's name should be "et al" or "and others" followed by a comma. Make sure you place a comma. For example, the data entered in the Author field of a reference as

Good Author Published Writer et al,	will appear in a bibliography as	[1] Author G, Writer P, et al.

But if you omitted the final comma from "et al", you will have inaccurate

Good Author Published Writer et al	will appear in a bibliography as	[1] Author G, Writer P, al e.

EndNote treats all the text before the comma as the last name. But if you don't enter the comma, then EndNote considers "al" as the last name and "et" as the first name.

- If a reference has no author, leave the Author field blank. Do not enter "Anonymous." EndNote can determine how to format anonymous works in a bibliography depending on the style of the paper. However, if a work is published with "Anonymous" as specified on the publication, then type "Anonymous" in the author field.
- To enter organizational author names, put a comma after the name. For example:
  National Heart Lung and Blood Institute,
  University of Minnesota,
  Make sure you do not insert commas in the name, because <u>*all the text before the comma is interpreted as last name, and after the comma as first name*</u>.
- In the above example, if you would like to include commas in the name of the organizational author such as "National Heart, Lung, and Blood Institute," the

trick is to replace the first comma with double commas and to not place the terminal comma. For example:
"National Heart,, Lung, and Blood Institute" will appear in bibliography
as
National Heart, Lung, and Blood Institute

- To enter complex names, put a comma after the entire last name, such as:
de Gaulle, Charles
This way both "de" and "Gaulle" are interpreted as last names.
- To enter suffix in an author name, enter the following: last name, first name, suffix. For example,
Smith, John, Jr.
Smith, John, III

> ⚠ **Alert:** If you have the "Suggest Terms as You Type" feature of the term lists enabled, this will suggest author names based on the assumption that the names are being entered in the "last name, first name" format (see next section for details).

Journal Names

- Enter a complete journal name—do not enter journal abbreviations. EndNote will create appropriate abbreviations automatically in the bibliography as specified by the output style.

Year

- Enter the year in full, such as "1992" or "2004."
- When appropriate, you may enter "In press" or "In preparation."

Pages

- Page range can be entered as complete (231–239) or as abbreviated (231–9). The number formatting in bibliography will depend on the output style and not how you entered it in reference.
- Do not use commas (e.g. "1,521) for page numbers in thousands. Instead just enter it as "1521."

Date

- Enter dates as you would like them to appear in manuscript. EndNote does not reformat dates during bibliography creation.

Term Lists

You should have a basic understanding of the Term Lists feature of EndNote to maximize accuracy and efficiency in the manual entry of references in EndNote.

What is a Term List?

Term List is a feature of EndNote that helps you perform faster and more accurate data entry into references. When you create a new EndNote library, EndNote automatically sets up three new empty term lists for you—Authors, Journals, and Keywords. As you enter new references into your library (by typing, importing, or pasting) EndNote updates the term lists automatically so that the Authors, Journals, and Keywords term lists include all of the author names, journal names, and keywords entered into your references.

Two Basic Features of Term Lists

(A) "Auto-Completion" or "Suggest Terms As You Type." This speeds up data entry. If you are entering text into a field that is linked to a term list, EndNote finds the first matching term in the list and suggests that as the term you want to enter. The suggested text appears highlighted after the cursor. Continue typing until EndNote suggests the correct term and then press ENTER or TAB to accept it. For example, in Figure 8, as you begin entering an author's name; EndNote attempts to complete the name for you by suggesting the closest matching name in the Authors term list.

(B) "Auto-Update." EndNote automatically updates the term lists in a library so they stay current with the data that has been entered into references. If you enter a new name that is not already in the associated term list, the term appears in red colored text to indicate that it is a new addition to the term list. When you close or save the reference, that new term is added to the list.

Journal Term Lists

EndNote includes predefined term lists for journals pertaining to three scientific fields: Medicine (Index Medicus), Chemistry, and Humanities. However, when you create a library, these terms lists are not automatically associated with the

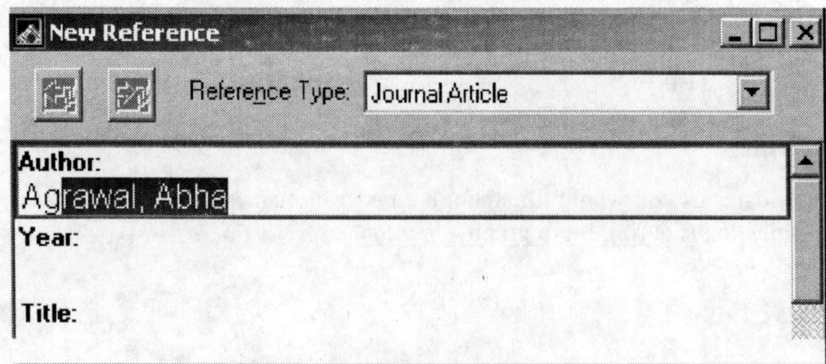

FIGURE 8. Auto-Completion feature of Term Lists

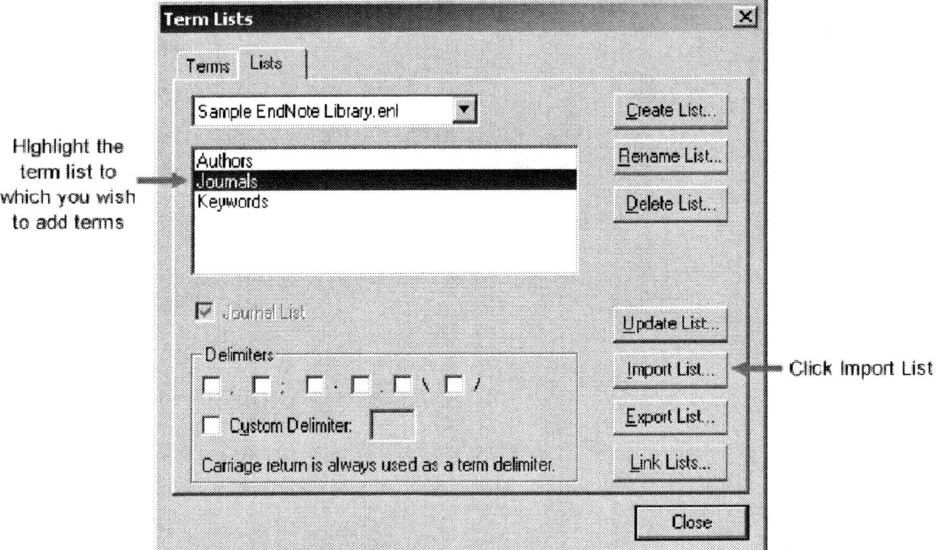

FIGURE 9. Selecting Define Term Lists

library. If you would like to use these term lists, you need to perform the following steps to import these terms lists into your library's Journals term list:

- Click *Tools > Define Term Lists* (Figure 9).
- A dialog box opens (Figure 10). Make sure the Lists tab is selected. Highlight the terms list (such as Journals) to which you want to add terms. Click *Import List.*
- In the next dialog box, locate the term list files by clicking the drop down menu. Generally, these term lists are stored in the C:\Program Files\EndNote\Term Lists folder. Select the text file to be imported and click *Open*.

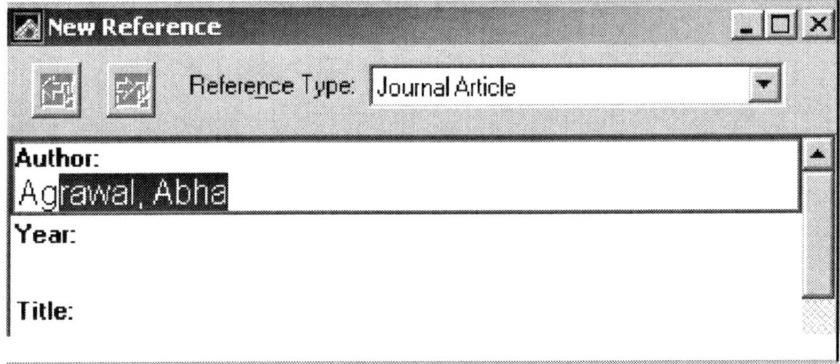

FIGURE 10. Importing a journal term list

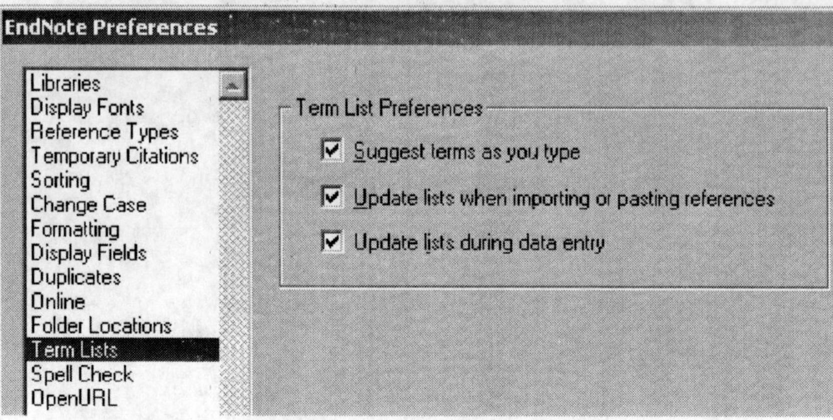

FIGURE 11. Setting Term List preferences

Technical Tip: The term lists are library-specific and associated to a particular library. Therefore, the import of predefined term lists must be done for each library that you want to use these terms with.

Helpful Hints About Term Lists

- Term lists are stored with the library and are specific to only that library.
- Up to 31 term lists can be created for any library.
- The number of terms in a term list is unlimited.
- Term lists are specific to a reference field. This means that if you enter text in the Author field, EndNote will check this text only against the terms in the Author term list (and not with terms in Keywords term list etc).

Turning Off Term List Features (Figure 11)

If you do not wish to use this feature of EndNote, you can turn it off. To do so:

- Click *Edit* > *Preferences.*
- Select Term Lists.
- Check or uncheck boxes according to your preference.

The 'Figure' and the 'Chart or Table' Type Reference

In addition to using EndNote for the usual text references such as journal articles and books, I find it very useful to use it for inserting pictures, charts, and tables in the manuscript. EndNote makes it easy for you to insert pictures or files in a paper, to label them with a caption, and to sequentially number them similar to text references.

'Figure' reference type is used to insert pictures and the 'Chart or Table' type to insert tables, chart, PowerPoint slides, and so on.

These two reference types are very similar. Both of these have a reference field called Image where you store your picture or table. One major difference between the two is how you enter data in the Image reference field. If you are in the Figure type, then you click *Reference > Insert Picture* command whereas for the Chart or Table type reference, you will click *Reference > Insert Object* command. Note that even in the Chart or Table reference type, you insert your file in an Image reference field (there is no chart or table field).

Creating a Figure Type Reference

- Click *Reference > New Reference.*
- In the new reference window, select Figure from the dropdown list.
- Type data in various reference fields as necessary.
- Type a Caption for the image in the Caption field.
- Insert an image.
  - Bring your cursor to the Image field in the reference.
  - Click on the *Reference > Insert Picture* (Figure 12).
  - In the next dialog box, locate the image you would like to insert and click *Open.*

Creating a Chart or Table Type Reference

- Click *Reference > New Reference.*
- In the new reference window, select Chart or Table from the dropdown list.

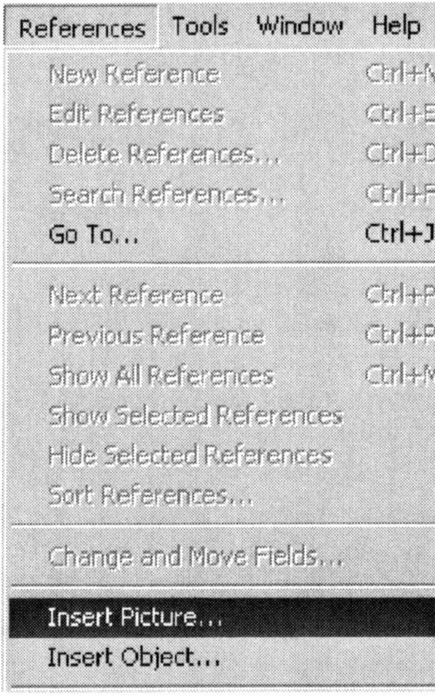

FIGURE 12. The Insert Picture command

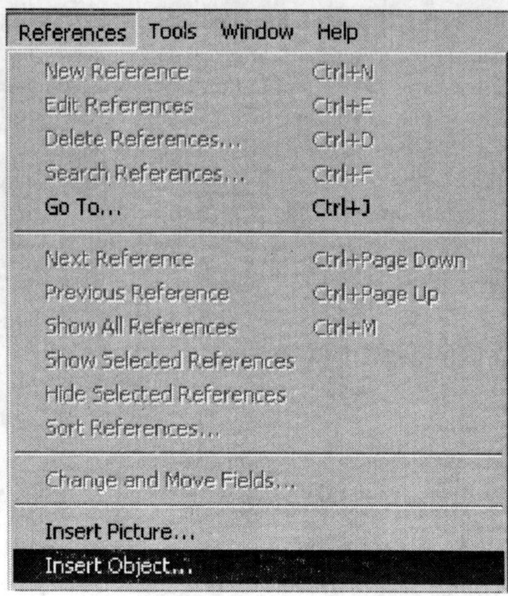

FIGURE 13. Insert Object command

- Type data in various reference fields as necessary.
- Type a Caption for the image in the Caption field.
- Insert the chart or table.
  - Bring your cursor to the Image field in the reference.
  - Click on the *Reference > Insert Object* (Figure 13).
  - In the next dialog box, locate the file you would like to insert and click *Open*.

Notes About Figure and Chart/Table Type References

- The inserted image in the Figure type reference appears as a thumbnail within the reference (Figure 14). Double-clicking on the thumbnail will launch the associated application for image viewing and editing.
- The inserted file in the Chart/Table type appears as an attachment, with the program icon (Figure 15).
- EndNote 7 supports *at least* the following image file formats:
  - Windows Bitmap—BMP
  - Graphic Interchange Format—GIF
  - JPEG File Interchange Format
  - Portable Network Graphic—PNG
  - Tag Image File Format—TIFF
  - If you are using Microsoft Image Composer, EndNote does not support the .mic format and you must save your image file in one of the supported formats.

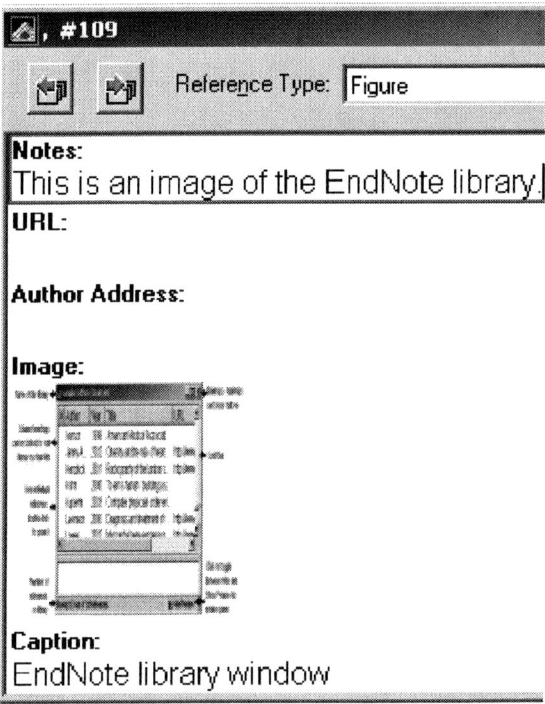

FIGURE 14. Figure type reference—thumbnail of image

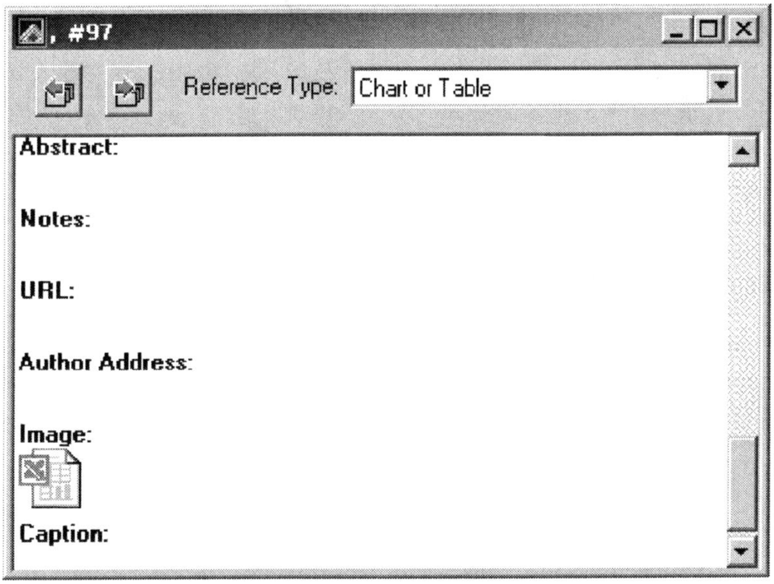

, #97
Reference Type: Chart or Table
Abstract:
Notes:
URL:
Author Address:
Image:
Caption:

FIGURE 15. Chart/Table type reference: inserted file as the program icon

- EndNote 7 supports *at least* these file formats:
  - Microsoft Excel, Access, PowerPoint, Word
  - Microsoft Visio
  - Microsoft Project
  - Text files (.TXT, RTF, HTML)
  - PDF files
  - Audio files (WAV, MP3)
  - Multimedia files (MOV, QuickTime)
- When you insert an image or file in a reference, EndNote places a copy of the image in a "*.DATA*" folder. Images in this folder are specifically linked to individual EndNote references. See Chapter 3 for more details about the "*.DATA*" folder.

Alert: Simply putting an image or file into the "*.DATA*" folder does not link it to a reference. You must use the *Insert Picture* or *Insert Object* command for this link to work properly.

Technical Tips:

- Each EndNote reference can contain only one image or file attachment. If you insert a second image or file, it will *replace* the first one.
- The image or file must be inserted under the *Image* field and not under the Name of File, Type of Image, or Image Source Program etc.
- The Caption field text determines the image or table caption when you insert it into a word document. The text in the Title field of the reference will define the title of the reference but will not determine the caption of the image/table.
- The Caption field accepts only simple text. If you type URL or file paths, they will be treated as simple text and will not create hyperlinks.

Entering Special Characters in References

EndNote supports entering special characters into references including diacritics, mathematical, and typographical symbols. To enter special characters, you can use one of the three methods:

A. Type the character on the keyboard, if the character is supported by the keyboard, such as % or & or $.

B. Copy the character from another program such as Character Map, a program supplied with Windows. To do this:
   - On your desktop, click *Program* > *Accessories* > *System Tools* > *Character Map* (Figure 16).

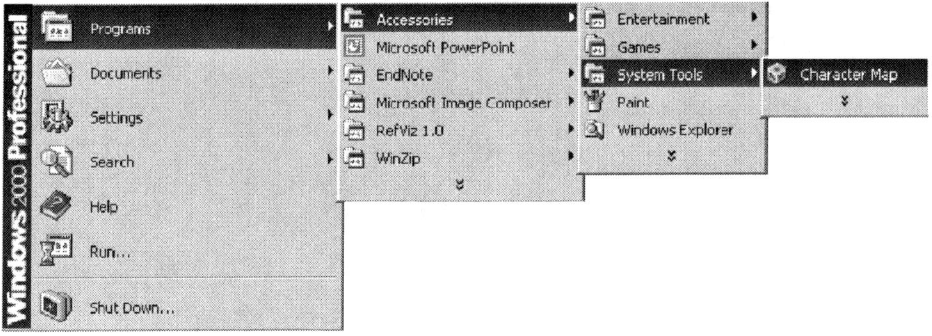

FIGURE 16. Opening the Character Map program

- A new window with the Character Map program opens (Figure 17).
- Click on the character you wish to insert, click on the *Select* button and then click *Copy.*
- Place the cursor in EndNote where you need to insert the character and click *Edit > Paste with Text Styles.* Note that doing *Edit > Paste* may not work properly and it is preferable to select *Paste with Text Styles.*

C. Type the ANSI or ASCII code for the special character on the **numeric keypad** of the keyboard while pressing the ALT key.

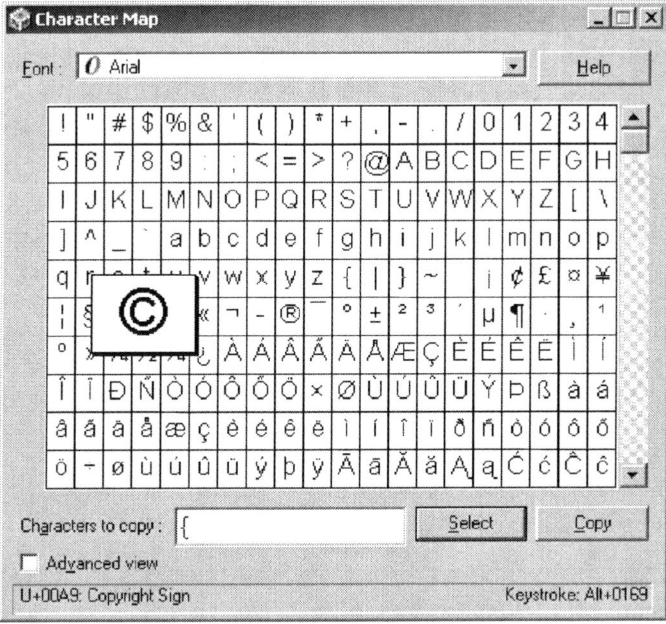

FIGURE 17. Character Map program

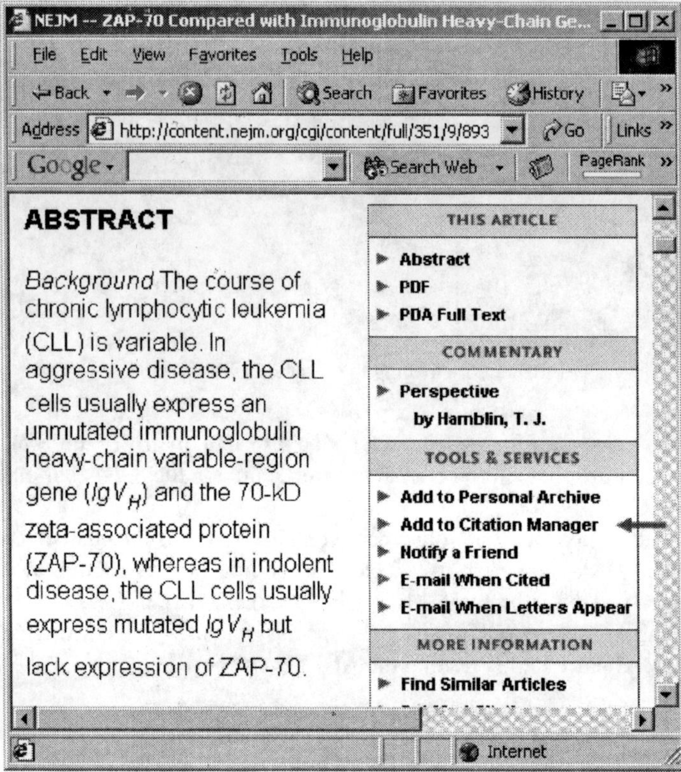

FIGURE 18. Selecting an article from a journal's website for download

Spell-Checking

While EndNote offers a spell-checker, you must have a reference open to perform a spell-check on it. You can not run a spell-check on the entire library with a single command. I find this to be a significant limitation of EndNote. Even if you open the maximum allowable ten references at a time, in a big library with lots of references, performing a check on one record at a time can be time-consuming and impractical, so you may want to run the spell-check at the completion of each reference entry.

To perform spell-check:

- Click *Tools* > *Spell-Check*. Note that the spell-check command is disabled if you don't have any open references.

> **Technical Tip:** EndNote's spell-checker works in all reference fields except the Author and the URL fields.

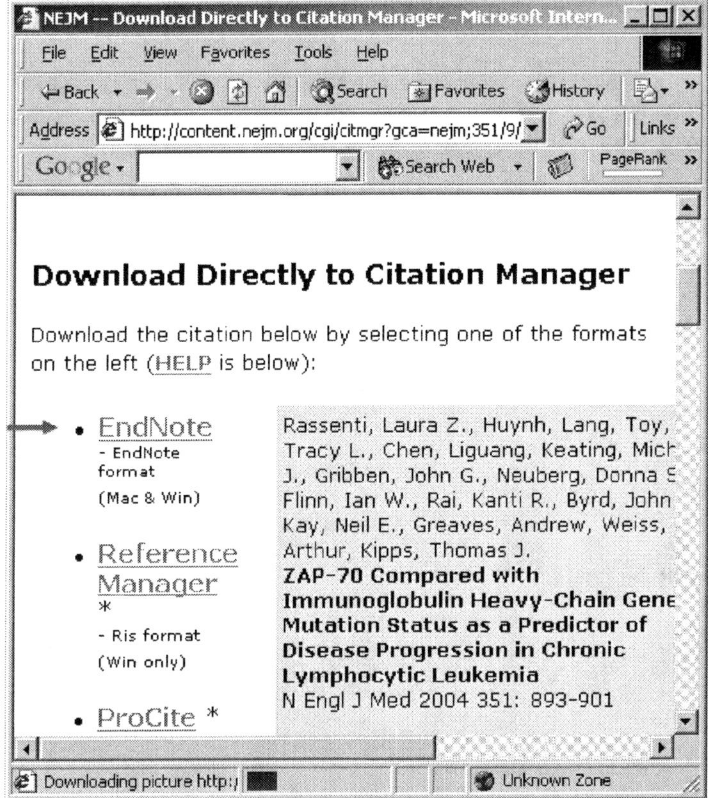

FIGURE 19. Downloading an online article—selecting EndNote

Downloading References from the Websites of Journals

Many biomedical journals are available online these days. If you come across an interesting article on the website of a journal that you would like to store for future reference, you can download the citation automatically into an EndNote library from the website. In the example below, I show you how to download a citation from the website of the New England Journal of Medicine (NEJM). The process works almost the same way for most of the journals' websites.

- Go to the NEJM's website by typing www.nejm.org in your browser.
- Locate the reference article that you would like to download. At this point, you should see a menu on the website that suggests a hyperlink to download this reference to a citation manager (Figure 18).
- Click on this link. In the next screen, click on EndNote (Figure 19).

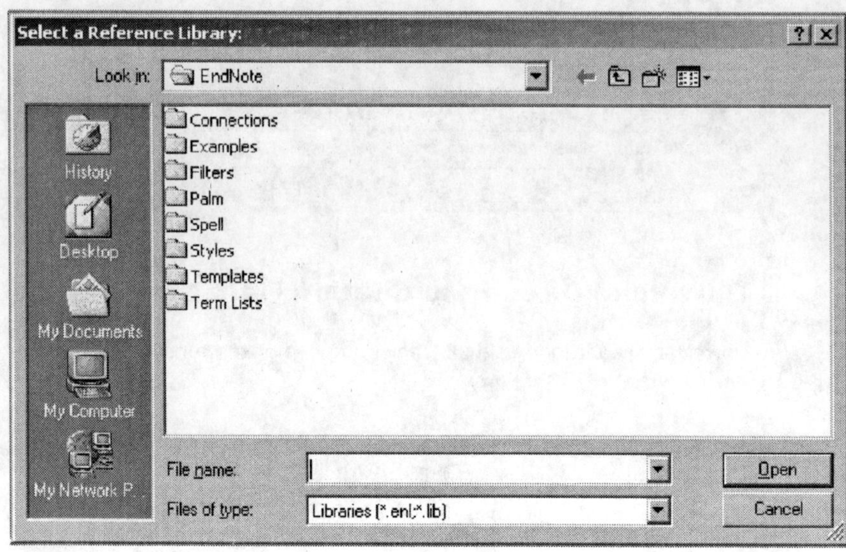

FIGURE 20. Downloading an online article—selecting an EndNote library

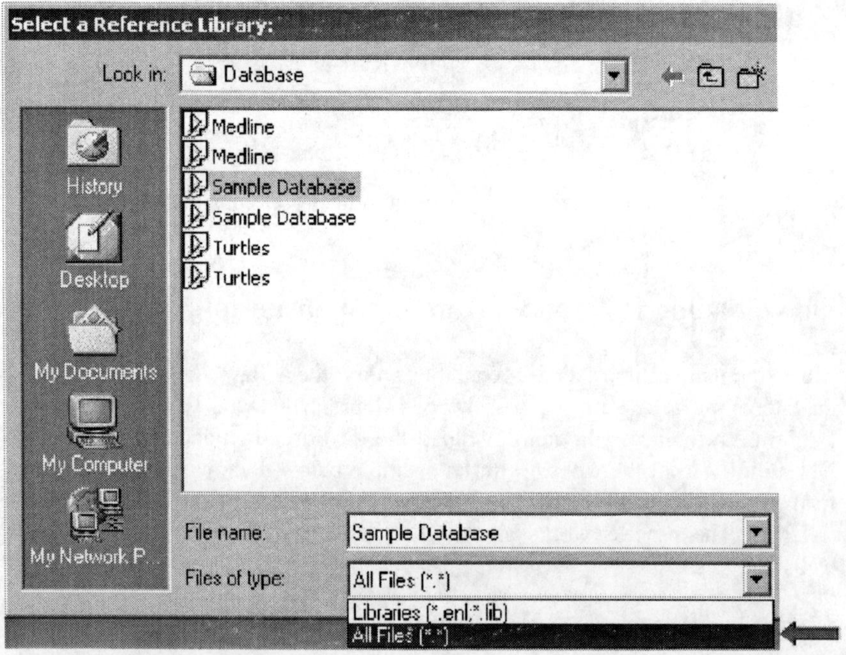

FIGURE 21. Opening Files Created by Programs other than EndNote

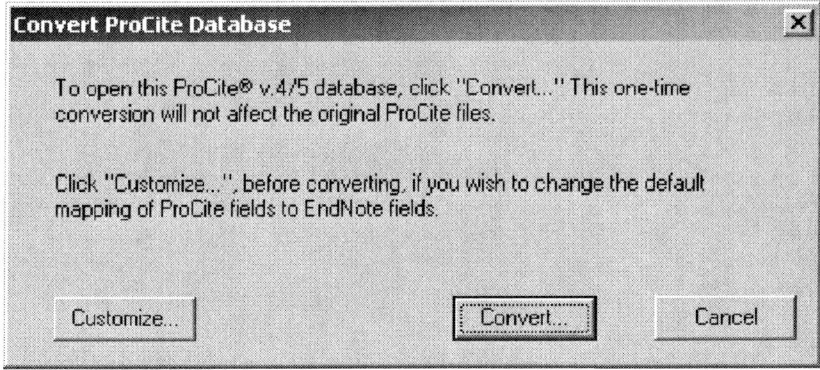

FIGURE 22. Conversion Dialog Box

- In the next screen, select the EndNote library where you would like this reference stored (Figure 20). Click *Open*
- The citation will be saved in the library. Note that you will only see the newly downloaded citation in the EndNote library window. Click *References > Show All References* in EndNote to see all references.

FIGURE 23. Customization during conversion

TABLE 2. Methods for importing references from other programs into EndNote

Program	Method for import
ProCite Reference Manager Refer/BiblX	Use File > Open command from EndNote
NoteBuilder® Papyrus 6 and 7® RefList® Ref 11® NoteBookII Plus® Citation 6 and 7® BibTex	Conversion utility available at EndNote's web site under the "Support and Services" section. URL: http://www.endnote.com/support/enconversion.asp
Bookends and Bookends Pro	Export from Bookends or Bookends Pro using the ProCite format. Then import resulting file into EndNote using the File>Open option.
Biblioscape Express	Export from Biblioscape Express using the 'EndNote refer' option. Then import the resulting text file into EndNote using File > Import.

Importing References from Other Reference Management Programs into EndNote

There are two methods to import references into EndNote from files created in a program other than EndNote.

(A) Files created in two major programs—ProCite and Reference Manager—can be directly opened by EndNote using the *File > Open* command without the need for any conversion. EndNote will do the conversion automatically. To perform this:
- In EndNote, click *File > Open.*
- Make sure you select *All files* under the "Files of type" option on the next screen (Figure 21).
- Click *Open.*
- In the next screen, you are presented with "Conversion" dialog box (Figure 22). You may choose to click on *Customize* button to customize fields during conversion—generally recommended for advanced users only (Figure 23). Click *Convert.*
- Select a name and location for the converted file. Click *Save*. Your file is now in the new EndNote format.

(B) For some other programs you may need to use a specific conversion utility. You can download conversion utilities for many programs from EndNote's website.

See Table 2 for various methods of importing references from other programs into EndNote.

6
Managing References in an EndNote Library

Things You will Learn in This Chapter

- About understanding the reference window.
- How to select single and multiple references.
- How to open, close, save, delete, and revert references.
- How to show and hide references.
- How to export references.
- How to search references.
- How to find and delete duplicate references.
- How to group edit references.
- How to link references to files and websites.

Chapter 5 discussed in detail various methods of entering references in an EndNote library. Now that you have a number of references in the library, this chapter discusses how to efficiently work with the references, for example, searching for specific references in a library.

Understanding the Reference Window

Before you start working with references, it is useful to understand the reference window. Important features of the reference window are illustrated in Figure 1. Some highlights include:

- The title bar displays the first author's last name, the year, and the EndNote record number.
- Next to the Reference Type label is a dropdown list of all reference types. The type of the current reference is displayed in the text-box next to it.
- Various reference fields and their contents are displayed in the main body of the window.
- Previous and Next Reference buttons can be used to quickly navigate through references.

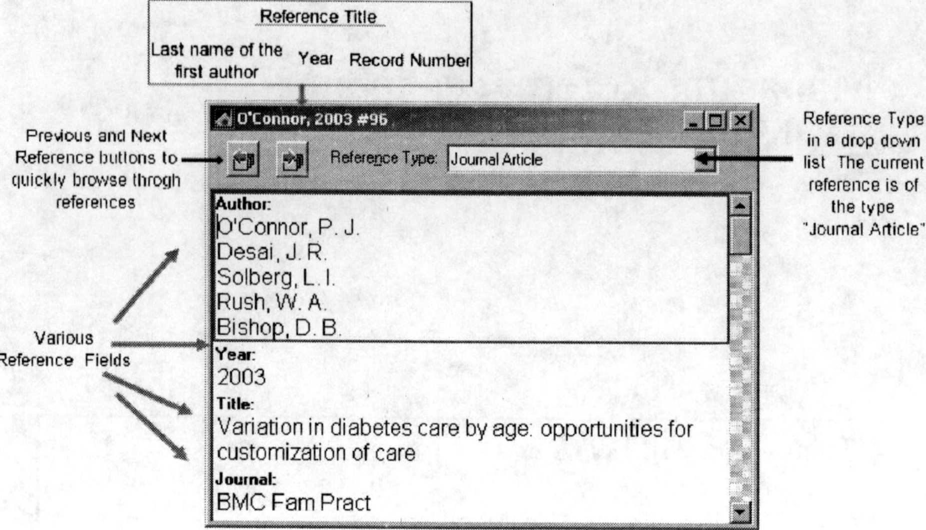

FIGURE 1. The Reference Window

Working with References

Selecting References

To work with an individual reference (such as open, save, or edit the reference), you will need to select it first. At other times, you will want to select multiple references, such as for copying, dragging, and so forth.

To Select a Single Reference, do one of the Following:

(a) Click on a reference to highlight it
OR

(b) Type the first few letters found in the field by which your EndNote library is currently sorted (see Chapter 4 for details). For example, if the library is sorted by the author field, you can type the first few letters of an author's last name to select the first reference by that author.

To Select Multiple References, do one of the Following:

(a) Hold down the CTRL key while clicking individual references (Figure 2)
OR

(b) To select a range of references, click on the first reference, then press the SHIFT key and click the last reference (Figure 3)
OR

(c) Click *Edit* > *Select All* to select all of the references showing in the library window (Figure 4). When all the references are selected, the *Select All* command changes to *Unselect All*.

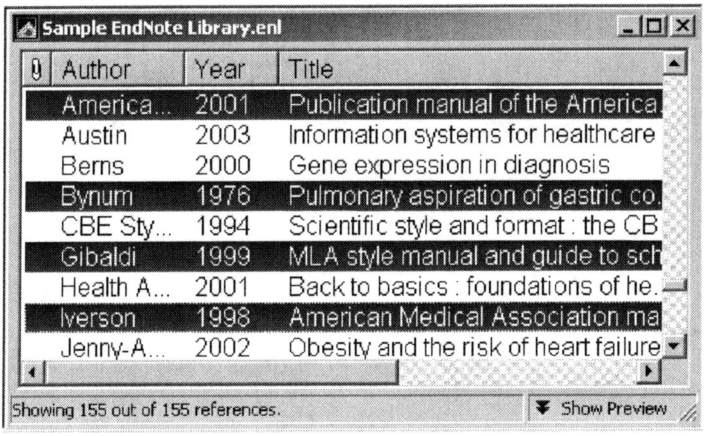

FIGURE 2. Holding CTRL to select multiple references

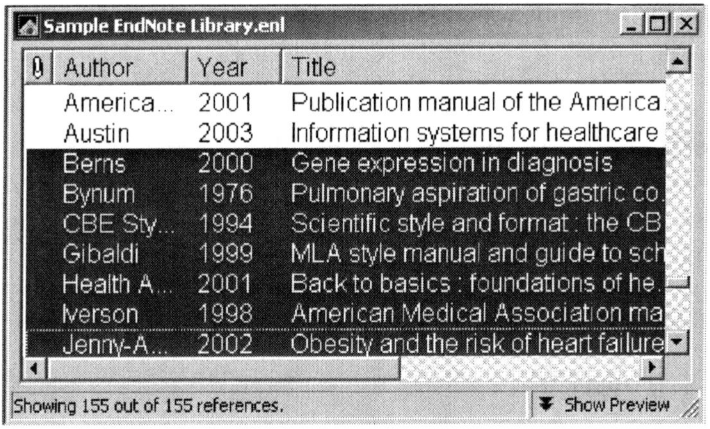

FIGURE 3. Pressing SHIFT to select multiple references

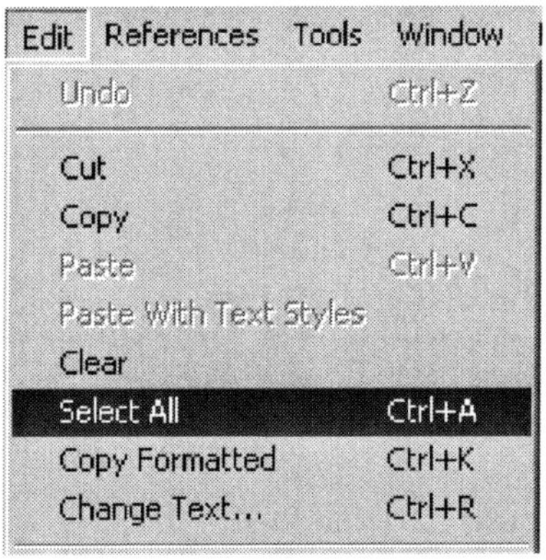

FIGURE 4. Edit > Select All to select all references

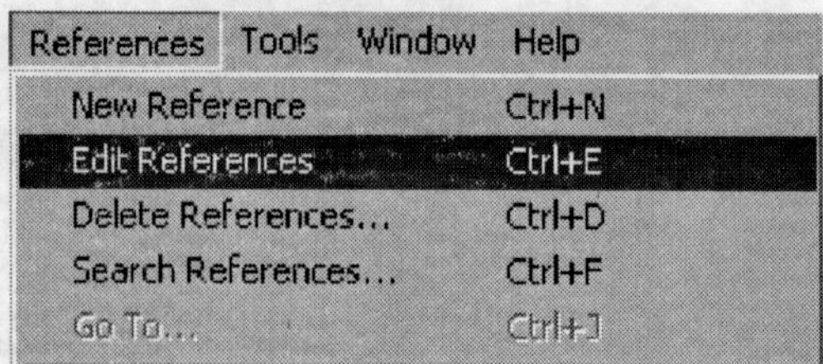

FIGURE 5. Selecting Edit References to open a reference

Opening References

First select the reference(s). Then do one of the following to open the reference(s):

(a) Double-click the selected reference(s)
OR
(b) Press *Enter*
OR
(c) Click *References > Edit References* from the menu (Figure 5).

> **Technical Tip:** A maximum of 10 references can be open at one time. The reference window that opens for each reference is where you enter and edit the reference data.

Simply click the "X" button in the right upper corner of the reference window to close the reference.

Saving References

To save a reference that you have created or edited, do one of the following:

(a) Click *File > Save* from the menu
OR
(b) Simply close the reference.

> **Technical Tip:** If you close the reference window or exit from EndNote while Reference windows are open, the references are automatically saved so you don't lose your work.

Deleting References

To delete reference(s) from a library, select the reference(s) and do one of the following:

(a) Click *Edit > Clear*
OR
(b) Click *Edit > Cut*
OR
(c) Click *References > Delete References*.

⚠ **Alert:** If you delete any reference from the library, the record number associated with the reference is deleted as well. Even if you paste this reference back into the library, this reference is assigned a new number. If you have any papers that use this reference, you may not be able to reformat that paper because the citation in the paper would be associated with the old record number. My suggestion is that you do not delete any references from your library that may be linked to a paper.

Reverting References

This command discards all changes made to a reference since it was last opened or saved. To do this, click *File > Revert Reference*. This command is not available once you close the reference.

Showing and Hiding References

You can choose to display only a subset of references in the EndNote window. This is useful when you want to perform some operations en block only on some selected references such as:

- Performing a search only on a subset of references.
- Use the *Change Text*, *Change Field*, and *Move Field* commands to modify only the showing references.
- *Print* or *Export* a subset of references.

To show only selected references in the library window

- Select the references you want to show.
- Click *References > Show Selected References* menu (Figure 6).
- Click *References > Show All References* to revert back to seeing all references.

To hide selected references

- Select the references you want to hide.
- Click *References > Hide Selected References* menu (Figure 6).
- Click *References > Show All References* to revert back to seeing all references.

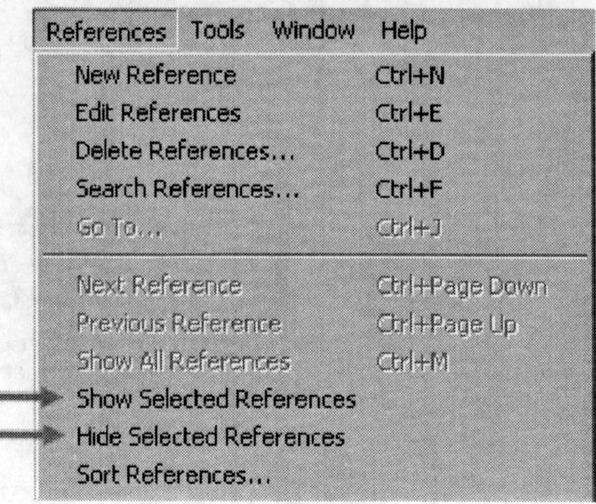

FIGURE 6. Show/Hide selected references

Exporting References

You can export references from an EndNote library to the following types of documents: RTF (Rich Text Format), HTML, XML, and plain text (.txt). This feature is useful for:

- Creating files that can be posted on the web (e.g., HTML files). See Chapter 4 for a detailed discussion of posting an EndNote library on the web.
- Creating files that can be imported into other databases (e.g., RTF or text files).
- Creating independent bibliographies (see Chapter 8).

Perform the following steps to export references:

- Make sure only the references you wish to export are showing in the library window.
- Select the desired order by clicking *References > Sort References*.
- Select the desired output style by clicking *Edit > Output Styles*.
- Click *File > Export*.
- In the next dialog box, type a name for the file and select a file format. Also choose the location where you want to save this file. Click *Save*.

Note the following points about exporting references:

- This command will export only the references that are showing in the EndNote library window.
- The formatting of references in the exported file will depend on the currently selected output style. See Chapter 8 for a detailed discussion of output styles.
- References in the exported file will appear in the order they are sorted in the library window.

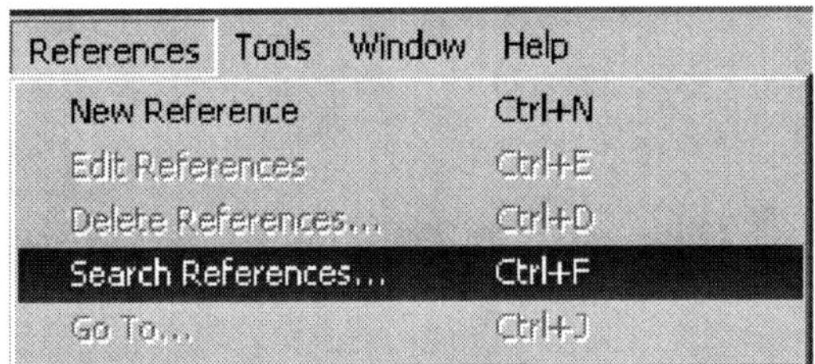

FIGURE 7. Launching EndNote search

> **Alert:** If your exported citations include images or files, note that the actual images or files are NOT exported regardless of the output style you choose. The images simply appear as text names, for example, "Image: 1551820288ure Type citation.bmp".

Searching References

EndNote provides a powerful search tool to help you search for specific references. This search tool can be used to search references within an EndNote library as well as to search Internet databases such as PubMed, the Library of Congress, and Ovid. This chapter describes using the search tool for searching references within an EndNote library. The same concepts apply to searching online databases as well. See Chapter 7 to learn more about using EndNote to search Internet databases.

Launching EndNote Search

To begin a search, click *References > Search References* from the menu (Figure 7). A search window displaying two search items appears on the screen.

Understanding the Search Window (Figure 8)

The Search window gives you various fields and functions to specify parameters for your search. Important components of the search window and their functions include:

- **Search Item**—consists of the following:
  - **Fields List**—a dropdown menu where you define the field to be searched (e.g. Author, Title, Year).
  - **Comparison Operator**—defines the search criteria for the search term. See Table 1 for a list of comparison operator functions and examples.
  - **Search term box** —text box to type the search term.
- **Boolean Operators**—"And," "Or," and "Not" radio buttons between the search items specify how various search items are to be combined. See Figure 9 for a graphic display of Boolean operators.

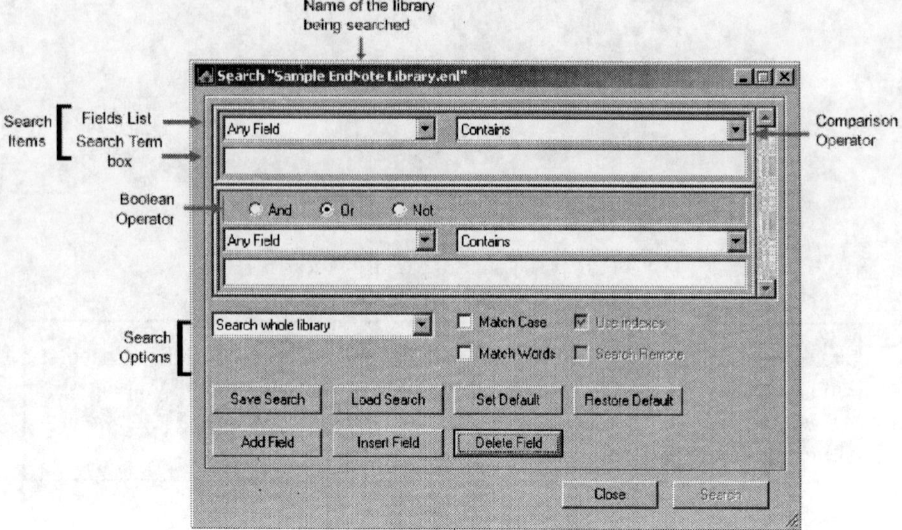

FIGURE 8. The Search window

And = References that match both search items. "And" narrows the search result.
Example: "Heart Failure" AND "Diabetes" searches only for references that mention both terms.

Or = References that match either of the search items. "Or" broadens the search result.
Example: "Heart Failure" OR "Diabetes" searches references that mention either term.

TABLE 1. Comparison operators

Field	Comparison operator	Search term	EndNote finds
Author	Contains	Smith	All references with Smith as an author
Author	Is less than or equal to	C	All references with authors whose last name begins with A, B, or C
Author	Is greater than or equal to	S	All references with authors whose last name begins with S–Z
Year	Is greater than	1995	All references published after 1995
Year	Is less than	1995	All references published before 1995
Title	Is less than	A	All references whose titles start with a number
Title	Contains	Heart	All references with "Heart" in the title
Title	Is	Heart	All references with "Heart" as the exact title
Keywords	Is		All references with no keywords

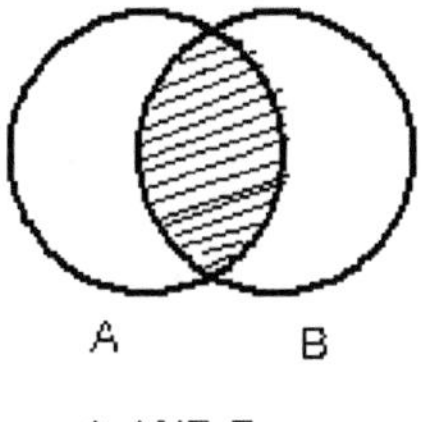

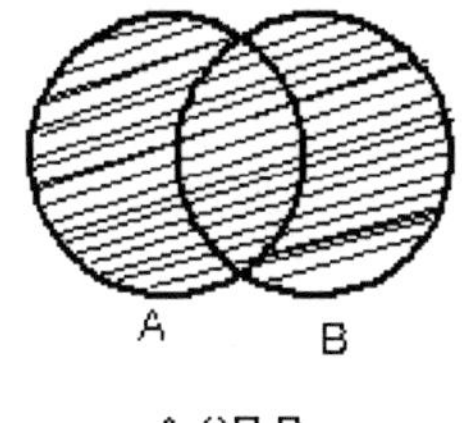

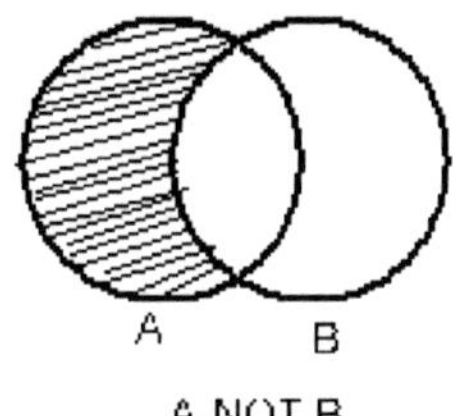

Circle A = Refernces containing "Diabetes"
Circle B = References containing "Heart Failure"
Shaded area = Search results using a given Boolean operator

FIGURE 9. Boolean operators

Not = References that contain the preceding term and not the later term. Example: "Heart Failure" NOT "Diabetes" finds articles that mention only heart failure and excludes any that mention diabetes.

- **Search options**
  - Search remote—must be checked for searching online databases.
  - Search set list—search whole library or omit from showing references.
  - Match case—by checking this box, you are instructing EndNote to match the search term exactly including its capitalization. For example, if you typed AIDS as the search term, by checking the "Match Case" box, EndNote will ignore any citations containing "Aids" or "aids" and will only find citations containing "AIDS."
  - Match Words—by checking this box, you are instructing EndNote to find only exact matches to the search words and to ignore partial matches.
- **Other functions**
  - Save search—You can save the configuration of the search window by clicking "Save Search." Note that the search configuration is saved as a file with ".enq" extension. You can run a saved search strategy later.
  - Load search—Click this button to load a previously saved search configuration. Saved searches also keep the search terms in them, making it easy to perform the search later.
  - Set default—use this button to save the configuration of all of the items in the search window **except for the search terms.** The default configuration will appear whenever you first open the search window.
  - Restore default—click this button to clear all search terms currently entered in the search window and to reset all settings to the default configuration.
  - Add fields—adds a new search item to the end of the list.
  - Insert fields—same as the "Add Fields" button except that it adds a new search item immediately before the selected search item.
  - Delete fields—remove the selected search item.

 **Technical Tip:**

- Pressing Esc cancels a search in progress.
- Diacritical marks (accents) can be used in searches. Letters such as e′, û match those letters exactly. So a search for "resûme" will not find "resume."

⚠ **Alert:** The search references command only searches one library at a time. If you have more than one library open, only the active library will be searched. To verify what library you are searching, always check the name of the library in the title bar of the search window (Figure 8).

Performing Search

- Enter search parameters in the search window, such as the text to be searched, in the Search Term box.
- Click *Search*.
- The next screen will show you the references that match your search criteria.

Duplicate References

Your EndNote library may contain duplicate references especially if you import references from multiple databases. EndNote provides an easy tool to search for duplicate references in a library so that you can delete them if needed. Note that this command applies only to references showing in the library; if you would like to search the entire library for duplicates, make sure to click *References > Show All References* menu.

Checking for Duplicate References

- Click *References > Show All References* menu.
- Click *References > Find Duplicates* (Figure 10).
- EndNote will highlight all duplicate references in the library (Figure 11).

 **Technicals Tip:**

- To look for duplicates, EndNote compares the Author, Year, and Title fields across the same reference type (such as Journal Article or Book).
- EndNote will not find every duplicate reference specially if the information in above fields is slightly different or if the reference types are different. Sorting the library by title can be another quick way of detecting duplicates visually if your library is not too big.

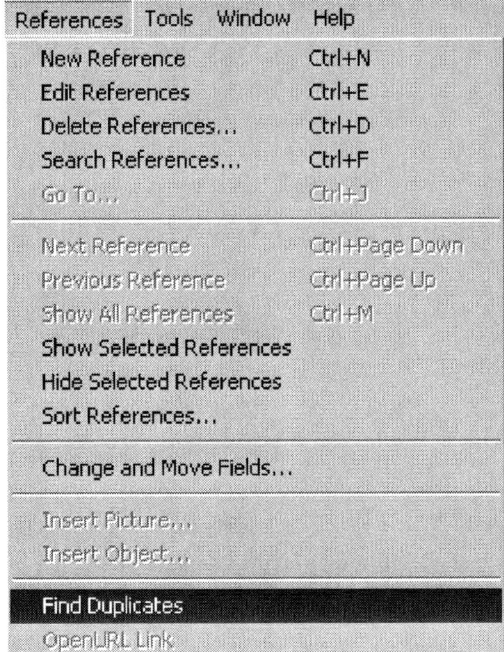

FIGURE 10. The Find Duplicates command

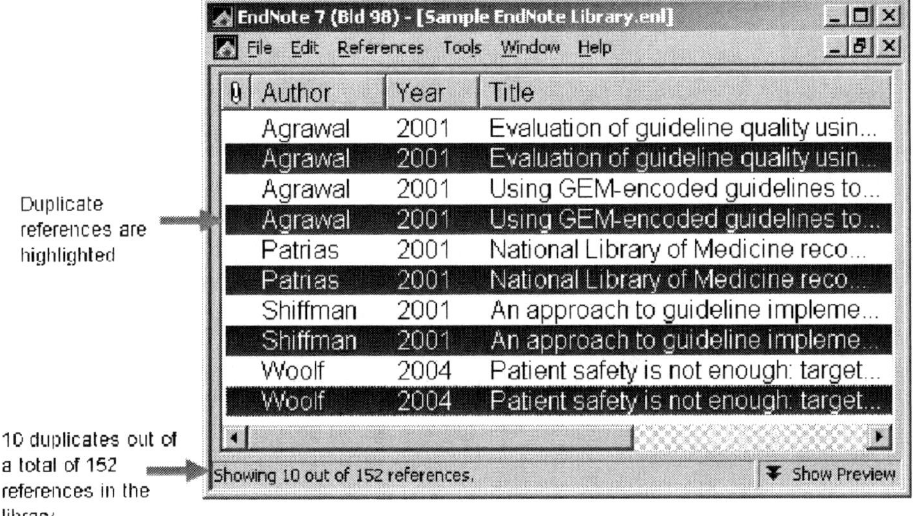

FIGURE 11. Finding duplicate references in an EndNote library

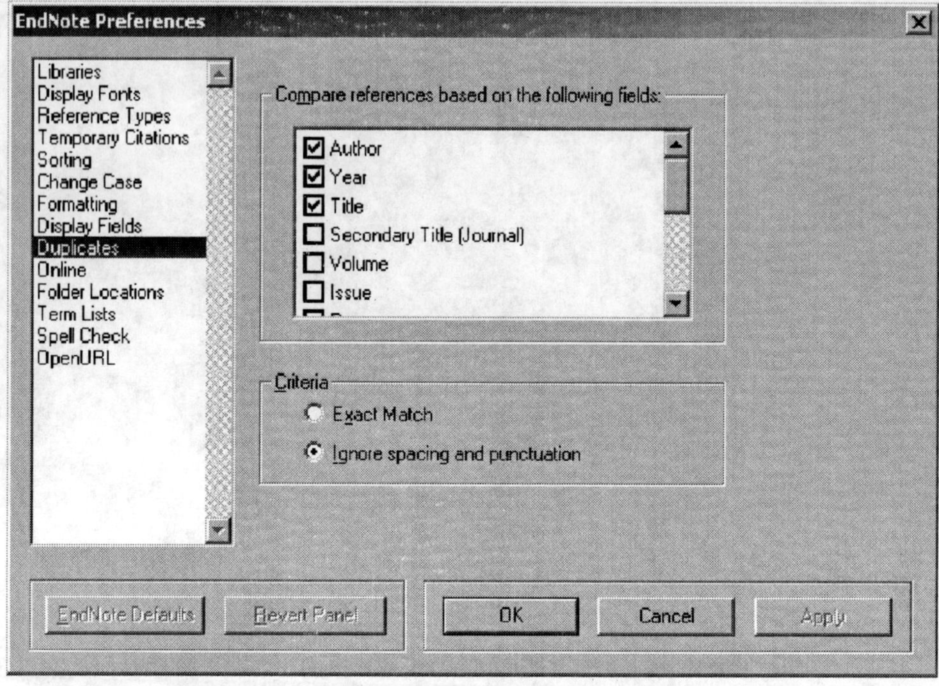

FIGURE 12. Customizing settings for Find Duplicates command

Customizing Settings for Find Duplicates Command

You can change the criteria for detecting duplicates by customizing EndNote. To customize settings for duplicates checking:

- Select *Edit > Preferences.*
- Click Duplicates (Figure 12).
- Click in check boxes to select fields by which you would like EndNote to check references. Select one of the boxes from the *Criteria* depending upon how exact you like your duplicates search results to be.

Technical Tip: Be careful that checking too many fields will make the search very stringent. Checking too few fields will cause the search to be non-specific.

Deleting Duplicate References

Technical Tip: Before deleting duplicate references, check the record number to make sure that you are not deleting any references already cited in a paper. If you do, then you won't be able to access that reference in your bibliography. An easy way to check the record numbers is to customize your library to display record numbers in the library window. To do this:

- Click *Edit > Preferences.*
- Under the Display Fields option, select Record Number as one of the columns to be displayed (Figure 13).

Group Editing of References

EndNote provides three editing commands to change data in a group of references so that the data in your library remains consistent (Table 2).

Alert: The changes made by these three commands can NOT be undone—the Undo command does not work after these operations. So make sure to back up your library before performing the group editing operations.

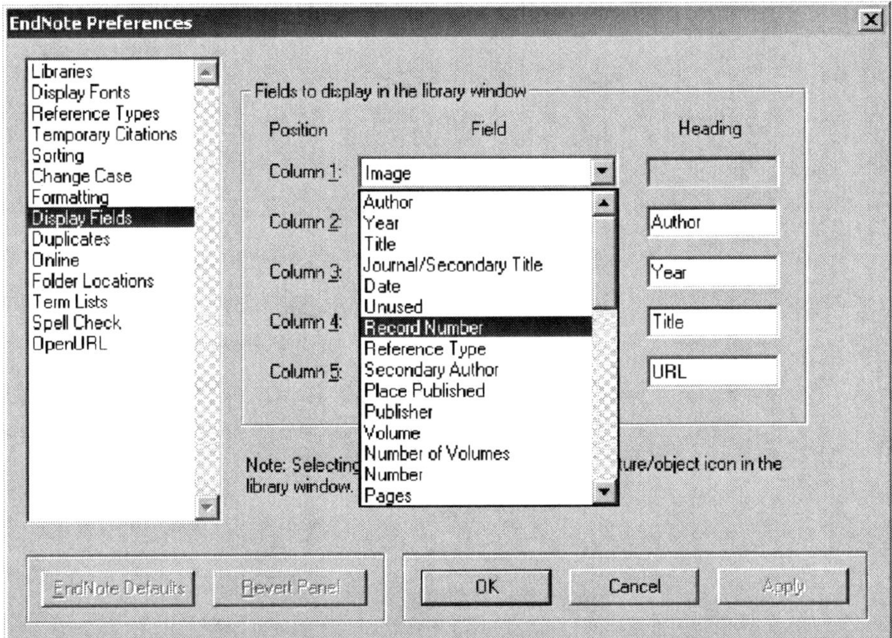

FIGURE 13. Editing EndNote preferences to display record number

TABLE 2. Group editing of references

Command	Description	Example usage
Change Text	Searches for text in your library, and either deletes that text or replaces it with other text that you specify.	Searching for misspelled term and replacing it with the correct spelling. Replacing abbreviations with full names or vice versa. Cleaning keywords to replace one term with a new one or to delete keywords you no longer want.
Change Field	Modifies any field in library by (a) Inserting text at the beginning or end of the field or (b) Replacing all contents of a field with different text or (c) Deleting the contents of the field	Add a unique keyword to a set of references to make later retrieval easier. Add the date or source of data to a batch of newly imported references.
Move Field	Moves the entire contents of a field to a different field.	Specially useful to correct any inaccuracies in imported references from online databases. For example, if your import places all information for the Note field in the Abstract field, you can use the Move Field command to easily move the information back to the Note field.

The Change Text Command

This command works like the "Find/Replace" command in Microsoft Word. To change/replace text in references:

- Select the references you would like to edit. Make sure only these references are showing in the library window.
- Click *Edit* > *Change Text* (Figure 14).

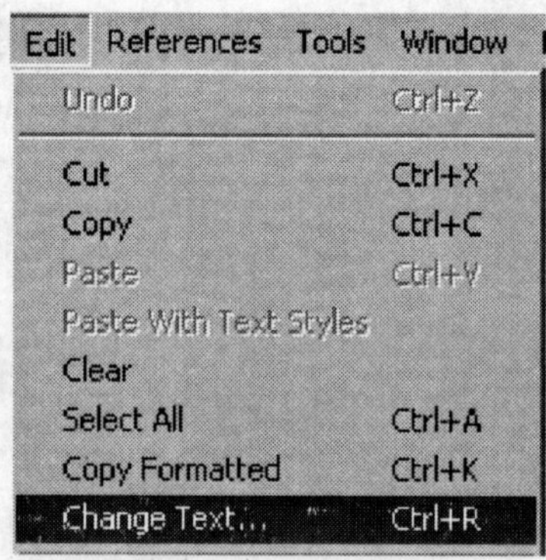

FIGURE 14. The Change Text command

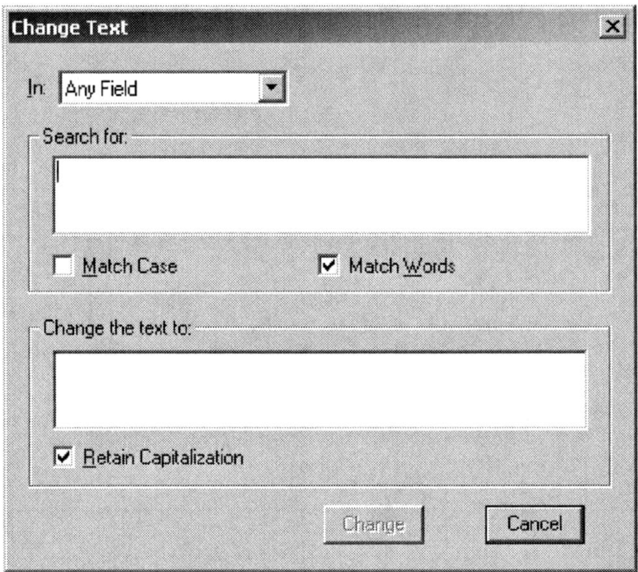

FIGURE 15. Change Text command options

- Fill in the options in the next screen (Figure 15).
  - Select the field to be searched from the field list.
  - Type the text to be changed in the "*Search for:*" box.
  - Change the "*Match Case*" or "*Match Words*" settings if necessary.
  - Type the text that should replace the original text in the "*Change the text to:*" box.
  - Select the "*Retain Capitalization*" option to maintain the same capitalization as the text being replaced.
- Click *Change* to execute the command.

The next screen presents a dialog box with the information about the changes. Click *OK* to confirm or *Cancel* to cancel the operation (Figure 16).

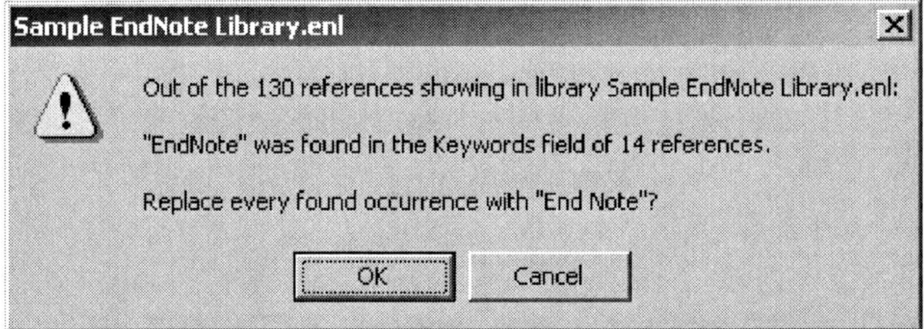

FIGURE 16. Change Text command warning

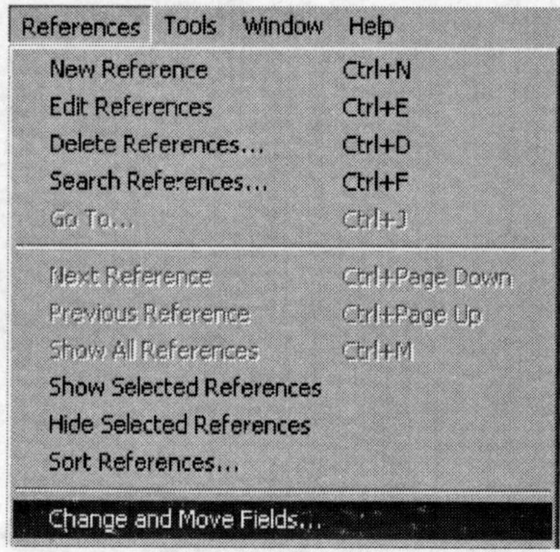

FIGURE 17. The Change and Move Fields command

Technical Tip: If you need to search for text and then delete it, simply leave the "*Change the text to*" box empty.

The Change Fields Command

This command also applies only to the references showing in the library window. To change the contents of a given field in all showing references:

- Click *References > Change and Move Fields* (Figure 17).
- Make sure *Change Field* tab is selected (Figure 18).
- From the *In field* list, select the field you want to modify.
- In the text box, enter the text that should be added to the field.
- Select one of the following options:
  - *Insert after field's text*—appends text at the end of the chosen field. **It does not modify the text already in the field**.
  - *Insert before field's text*—appends text at the end of the chosen field. **It does not modify the text already in the field**.
  - *Replace whole field with*—replaces the entire content of the field with the text entered in the text box.
  - *Clear field*: deletes the entire contents of the chosen field.
- Click *Change*. The next screen presents a dialog box with the information about the changes. Click *OK* to confirm or *Cancel* to cancel the operation.

The Move Fields Command

This command also applies only to the references showing in the library window. To change the contents of a given field in all showing references:

- Click *References > Change and Move Fields* (Figure 17).
- Click the *Move Fields* tab (Figure 19).

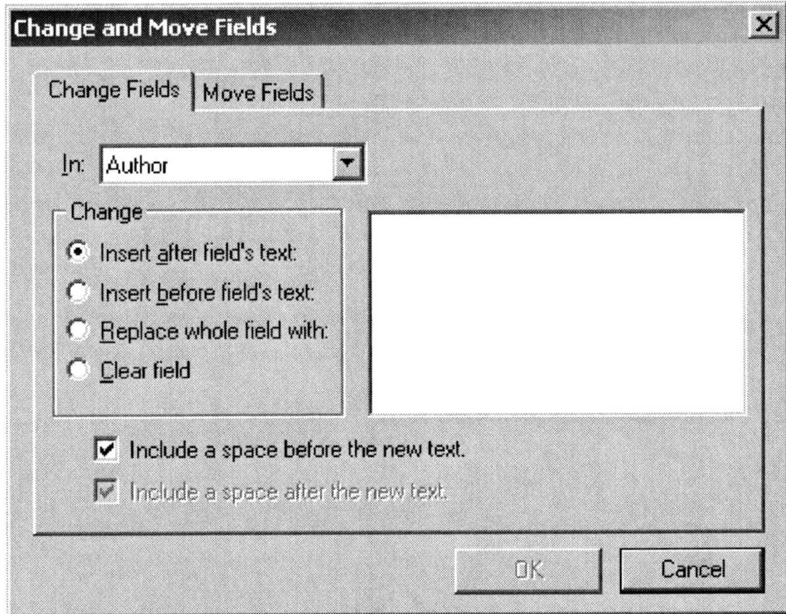

FIGURE 18. Change Fields options

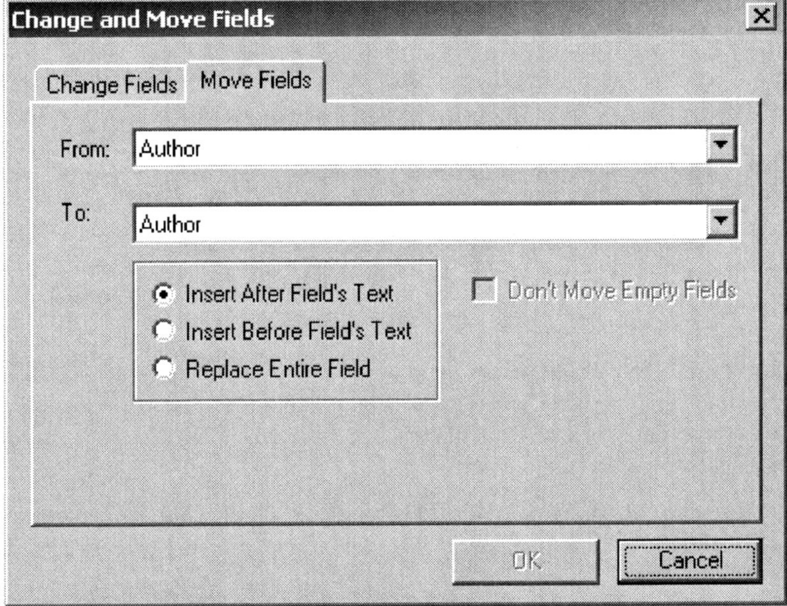

FIGURE 19. Move Field options

- Use the "*From*" list to select the field you want to move the data from.
- Use the "*To*" list to select the field to where you want that data to go.
- Select one of the three options by clicking on a radio button:
  - Insert moved text <u>after</u> the data in the destination field.
  - Insert the moved text <u>before</u> the data in the destination field.
  - Replace the content of the destination field with the moved text.
- Check or uncheck the "*Don't move empty fields*" box. If you uncheck it, then any reference with an empty originating field deletes the contents of the corresponding destination field. If this box is checked, the contents of the destination field will *not* be deleted if the originating field is empty.
- Click *OK*.

Linking References to Files and Websites

Linking References to Files

If you have a journal article stored as a reference in your EndNote library and also have a full-text copy or an abstract or a graphic related to the article stored in your computer, it would be useful to link the two. This will allow you to quickly access the related file when you are browsing this reference in the EndNote library.

EndNote provides the ***"Link To"*** command to link a reference to any file in your computer and the ***"Open Link"*** command to open that file.

> ⚠ **Alert:** You must have the necessary program installed on your computer in order for EndNote to open an external file. For example, if you establish a link to a full-text journal article in the PDF format, you must have Adobe Acrobat Reader available in your computer to be able to open the file using the "Open Link" command.

To Link a File to a Reference:

- Select a reference in the library window or open a reference.
- Click *Reference > Link To* command (Figure 20).
- In the next dialog box, select a file that you would like linked to the selected (or open) reference, and click *Open*.

This command enters the path (to access this file on **your** computer) and a hyperlink to the chosen file into the URL field of the reference (Figure 21).

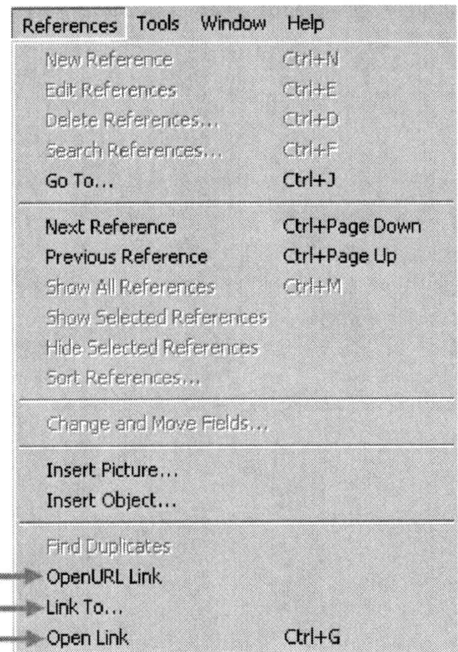

FIGURE 20. The Link To, Open Link, and OpenURL Link commands

Technical Tip: The *Link To* command is available only when a single reference is selected or open. If you select more than one reference, the *Link To* command is grayed out.

You can establish a link to more than one file in a single reference by clicking *Link To* again. This will create a path and hyperlink to the second file in the URL field of the reference. However, when you click *Open Link* command, it will open the first file only. If you want to open the second file, open the reference, go to the URL field and click on the hyperlink to the second file.

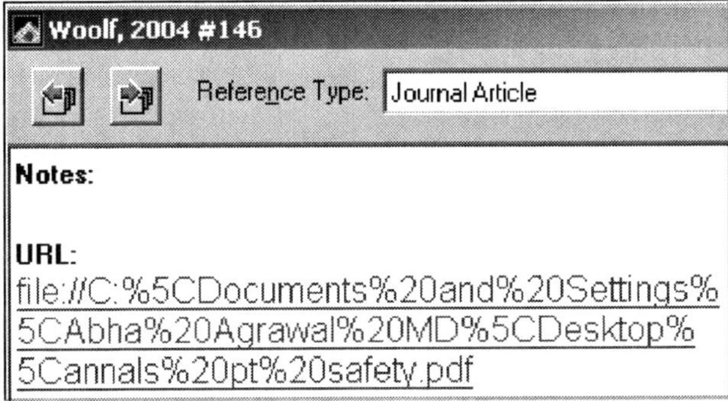

FIGURE 21. Link in the URL field

To open a linked file:

- Select or open a reference in the library window.
- Click *Reference > Open Link* (Figure 20).
- The linked file opens with the appropriate program in a separate window.

Technical Tip: When multiple references are selected, EndNote checks the URL field of each reference and opens the first link it finds.

If a reference window is open and you have selected some text in the window. Now you click the Open Link command. In this situation, EndNote will not go to the URL field to look for a hyperlink; instead it will open a Web browser window (such as Internet Explorer) with the selected text as the URL.

Alert: Unlike a URL, which points to a file with a location on the web, the link to a file points to a file on your computer locally. If you are using this library on another computer (or change the location of the file on your computer), then this link will become unusable. If you anticipate using your library on different computers, it may be better to attach the file to a reference by using the *References > Insert Object* function. Doing so essentially places a copy of the file in the "*.DATA*" folder (see Chapter 4 for details) making it easy to move the file with the library from one computer to another.

Linking References to Websites

To link a reference to a website

Simply type the URL of the related website in the URL field of a given reference.

To open link to a website

(a) Open the reference and click on the URL directly OR
(b) Click *Reference > Open Link* to open up a web browser window with the web address.

OpenURL Link Command

The URL established in the OpenURL Link is not specific to a reference; rather this applies globally to the library. This URL will open for any reference whenever you click the OpenURL Link command.

Technical Tip: Note that the *OpenURL Link* command is distinct from the *Open Link* command. The *OpenURL Link* command has no relation with the URL field of a reference or to the *Open Link* command under the *References* menu.

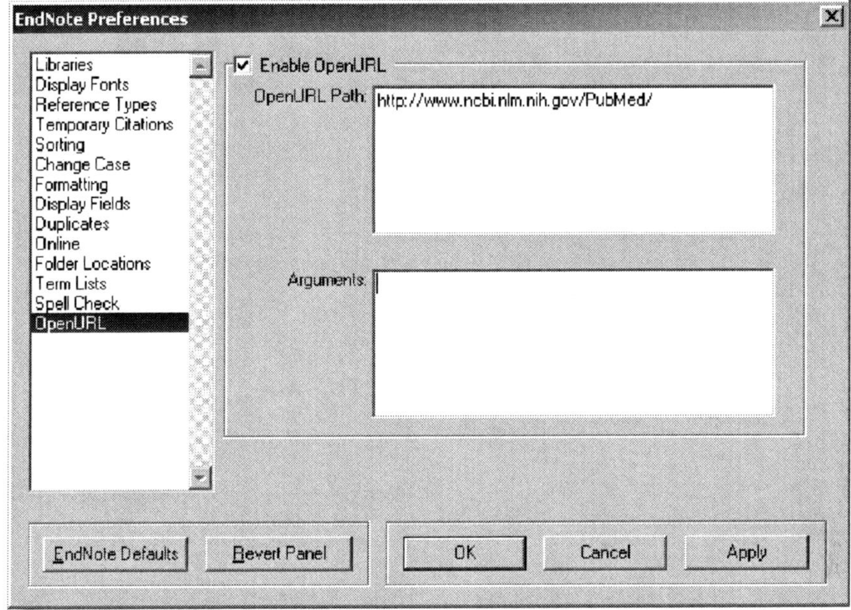

FIGURE 22. Enabling OpenURL Link command

This command is available only if you enable it by setting EndNote preferences. To enable and to set up a URL for this command (Figure 22):

- Click *Edit > Preferences*.
- Select *OpenURL* option.
- Check the *Enable OpenURL* box.
- Type the URL which you would like to open when you click the *OpenURL Link* command. In this example, I have typed the URL for PubMed so I can easily find articles related to a reference.

To open this link:

- Click *References > OpenURL Link* (Figure 20). A new browser window with PubMed will appear on your screen.

7
Using EndNote with Internet Databases

Things You will Learn in This Chapter

- Various methods of connecting EndNote with Internet databases and choosing the right connection method.
  - The Connection File method.
  - The Import Filter method.
  - The Direct Export method.
- How to get references from PubMed® into EndNote.
- How to get references from Ovid® into EndNote.
- How to get references from the Web of Science® into EndNote.

Introduction

The Internet is one of the most important sources of information for professionals in this information age. The Internet provides quick and convenient access to a wide variety of biomedical databases and journals.

Table 1 lists some Internet databases routinely used by healthcare and biomedical professionals.

After you have searched for references in one of the online databases, you will need a quick, convenient, reliable, and at least semi-automatic method to store selected references in a library. EndNote offers several user-friendly ways both to search references as well as to automatically retrieve them from an online database into an EndNote library.

My suggestion is to enter as few references manually into an EndNote library as possible. The process of manual entry is time-consuming, tedious, requires typing, and is prone to errors. While you may have to invest some time in learning automatic retrieval from online databases into EndNote, the benefits, some of which are described below, will more than compensate for your time and effort:

TABLE 1. Commonly used biomedical Internet databases

Name	Description	Provided by	URL
	Databases that do not require a subscription account		
PubMed	MEDLINE and PreMEDLINE databases	National Library of Medicine	http://www.ncbi.nlm.nih.gov/entrez/query.fcgi?db=PubMed
Library of Congress	Database of books containing approximately 12 million bibliographic records	Library of Congress	http://www.loc.gov/
USPTO	Database of US patents	United States Patent Office	http://www.uspto.gov
	Databases that require a a subscription account		
Web of Science	Web of Science, Web of Science AHCI, Web of Science SCI	Thomson ISI	http://www.isinet.com/products/citation/wos/
OVID	AIDSLine, BioethicsLine, Medline, Premedline, CINAHL etc.	OVID	http://www.ovid.com

- Using the automatic process is faster and more accurate.
- Reference details, such as author names, are always copied correctly without typographical errors.
- Journal names are automatically entered in the correct National Library of Medicine format.
- In addition to the routine details of a journal article (such as title, author(s) name, date, and journal), EndNote also retrieves and automatically fills in other sections of a reference, such as the URL, the abstract, and Keywords associated with the article (Figure 1). Keywords can be very useful for searching your EndNote library for specific references. You can click on the URL in an EndNote library to directly go to the article on the Internet. A growing number of journals give you the option of retrieving full-text articles from the web site, known as PubMed Central. Table 2 gives you the names of a few of those journals; a full list is available at http://www.pubmedcentral.nih.gov/index.html.

Various Methods of using EndNote with Internet Databases

There are three methods to search and retrieve references from an Internet database into EndNote:

(A) Connection files. In this method, EndNote makes a direct connection to the database so that you can use EndNote both to search the database and to retrieve references into a library.

(B) Import filters. In this method, you have to go outside of EndNote and search the database first (usually via your web browser) as you would if you were not using EndNote. Then you save the results as a text file and import this file

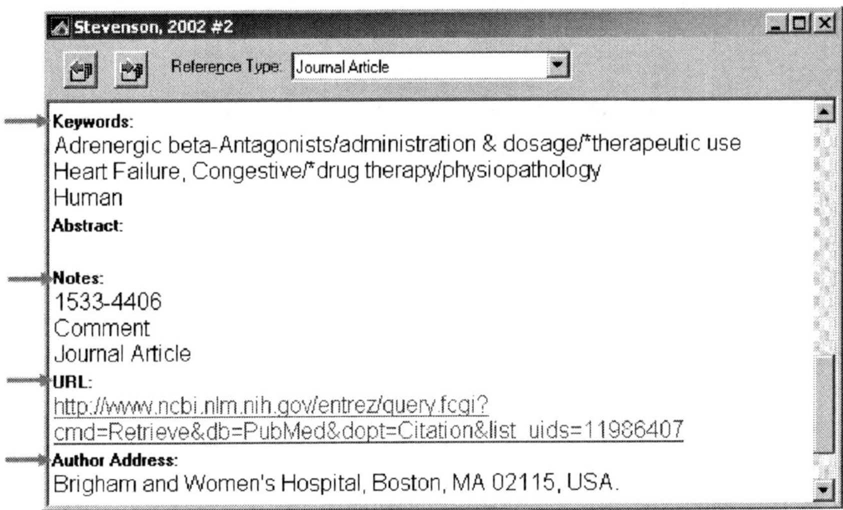

FIGURE 1. Additional fields imported by EndNote by automatic downloading of an article from PubMed

into EndNote using an **import filter**. Filters allow you to import citations into a library by telling EndNote how to interpret the information you have saved from a particular online database.

(C) Direct export. This is similar to the import filter method. Some databases have created a tool that automates the saving and filtering of citations; this method is called **direct export**.

TABLE 2. Some PubMedCentral journals

	Volumes in PubMedCentral		
Journal title	Latest	First	Access
Annals of Clinical Microbiology and Antimicrobials	v.3(1) 2004	v.1 2002	Immediate
BMC Cardiovascular Disorders	v.4(1) 2004	v.1 2001	Immediate
BMC Medical Informatics and Decision Making	v.4(1) 2004	v.1 2001	Immediate
CMAJ: Canadian Medical Association Journal	v.170(12) June 8, 2004	v.163 2000	Immediate
Health Research Policy and Systems	v.2(1) 2004	v.1 2003	Immediate
Journal of the American Medical Informatics Association	v.10(6) Nov 2003	v.1 1994	After 6 months
Journal of Clinical Investigation	v.113(12) June 15 2004	v.40 1961	Immediate
Molecular and Cellular Biology	v.24(11) June 1, 2004	v.1 1981	After 6 months

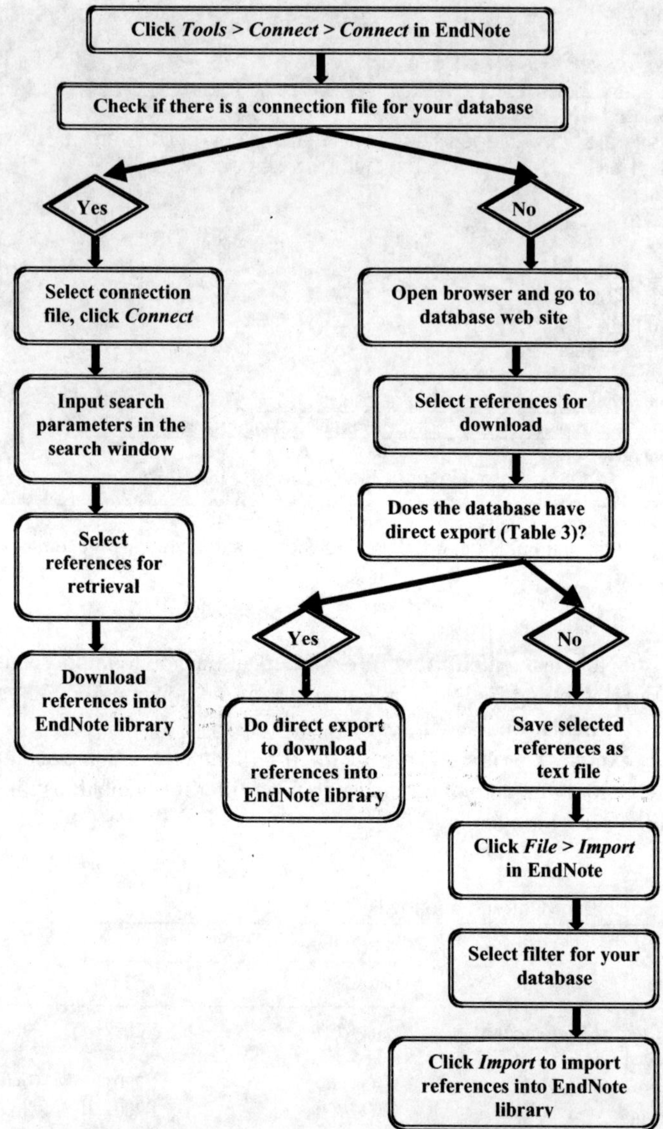

FIGURE 2. An overview of using EndNote with Internet databases

Figure 2 illustrates an overview of using EndNote with Internet databases.

Choosing the Right Connection Method

In general, the connection files method is the most direct and easiest way to get data into EndNote, because it involves no intermediate steps. However, to use the connection file method, the database you intend to use must be on a server that supports the Z39.50 protocol. See the next Technical Tip box for information about this protocol.

If a connection file for your database is not available, you can use the import filter method. The advantage of this method is that since you are searching a database on its own website, you have access to the full range of the search features, keywords, and thesauri of the database.

For connecting to a subscription database using connection files, you need to have an individual password for accessing the database. If you have access to a database such as OVID or the Web of Science through an organizational account (such as a university library) and do not have an individual password, you may not be able to access these databases using connection files. In that case, you can connect to these databases on the web, and then use the import filters or direct export to bring references into EndNote.

Many commonly used biomedical databases have a connection file or an import filter (or direct export feature) available to work with EndNote. The next sections of this chapter describe how to check which databases have connection files or import filters available for working with EndNote and how to download additional connection files and import filters.

Technical Tip: Z39.50 is an international standard for communication between computer systems, primarily library and information related systems. This standard protocol specifies data structures and interchange rules that allow a client machine (for example, your computer with the EndNote program) to search databases on a server machine (for example, the National Library of Medicine's PubMed server) and retrieve records that are identified as a result of such a search.

Z39.50 is becoming increasingly important to the future development and deployment of inter-linked library systems. The Z39.50 standard was originally proposed in 1984 to provide a standard way of interrogating bibliographic databases. It is maintained by the Z39.50 Maintenance Agency which is administered by the Library of Congress. For more information about this protocol, visit http://lcweb.loc.gov/z3950/agency/

Alert: You may experience difficulty connecting to a remote database via connection files if your computer is behind a firewall. This situation is not uncommon in large organizations. If so, you will get an error message "Host refused connection" or the connection will time out without connecting to the remote database. To fix this, you will need to contact your network administrator to open ports to allow communication through the firewall. Most (but not all) z39.50 connections use port 210.

The Connection File Method

What is a Connection File?

A connection file contains the information necessary to connect to, search, and import references from an online database into EndNote. EndNote needs a specific connection file for each online database such as the Library of Congress or PubMed.

What Connection Files do I Have?

The EndNote program has already installed on your computer hundreds of connection files for a variety of databases. These files are located in the Connections folder of the EndNote folder in the computer. The best way to view a list of available connection files is by opening the Connection Manager in EndNote by performing following steps:

- Click *Edit* > *Connection Files* > *Open Connection Manager* (Figure 3).
- The next screen is the Connection Manager Window displaying a list of various connection files (Figure 4).

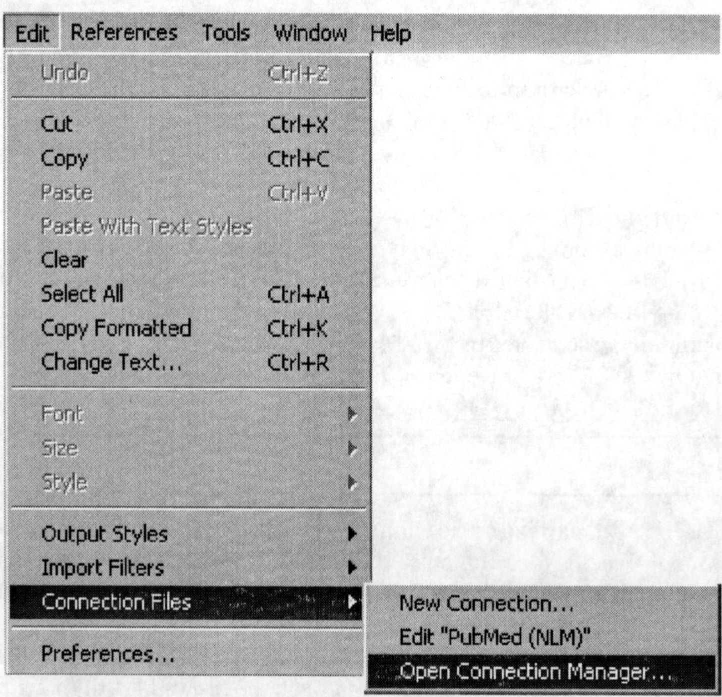

FIGURE 3. Opening the Connection Manager

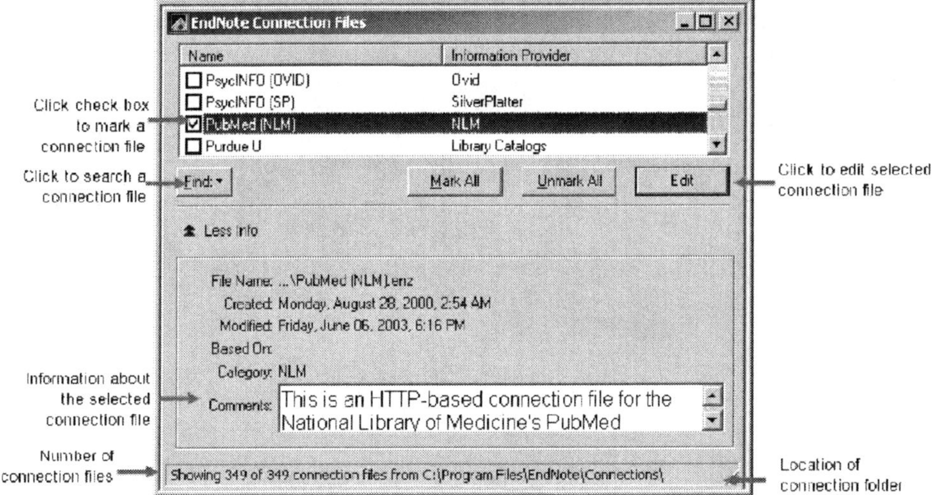

FIGURE 4. The Connection Manager window

Working with the Connection Manager

The Connection Manager window displays a list of connection files available in the Connections folder and gives you the options to edit them or select them as "favorites" for quick access when you use the "Connect" command. Figure 4 highlights some important features of the connection manager window.

Setting "Favorite" Connection Files

If you use a few databases frequently, it is a good idea to mark these connection files as "favorites" to save time each time you need to connect to the database. To perform this function:

- Click *Edit* > *Connection Files* > *Open Connection Manager* (Figure 3).
- In the Connection Manger, check the box next to the files you would like to set as favorites (Figure 4).
- You should now see these databases when you click *Tools* > *Connect* to connect to a database (Figure 5).

You can mark as many connection files as you like as favorite.

Downloading Connection Files from the Internet

If you don't find the connection file you are looking for in the connection manager, go to EndNote's web site where you will find additional connectional files available for download. Some online databases update their structure. This may make the connection file that came with the EndNote obsolete, and you may need to download a new connection file from EndNote's web site:

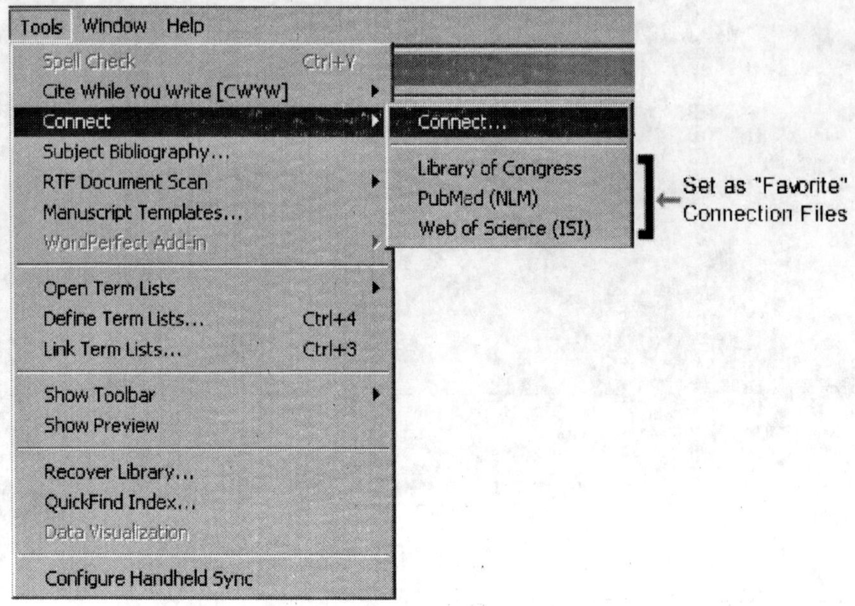

FIGURE 5. The connection file method

- Open your Internet browser and go to http://www.endnote.com/support/enconnections.asp.
- Select the connection file you would like to download and click on the "FTP" hyperlink (Figure 6).
- Click *Save* in the next dialog box to save this file to your computer (Figure 7).
- Save this file in the Connections folder. Most likely the path to this folder is C:\Program Files\EndNote\Connections, if you accepted the default options

FIGURE 6. Selecting a connection file for download from EndNote's web-site

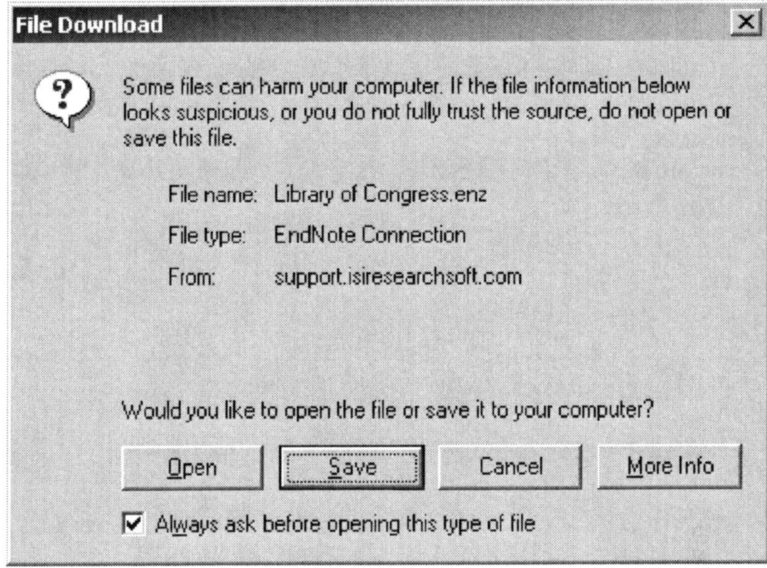

FIGURE 7. Saving a connection file to your computer

during installation. If you can't locate the folder, go to the Connection Manager window and the path to the folder is displayed in the bottom status bar of this window (Figure 4).

- If you already have an older version of the connection file, you will receive a 'replace' warning. Click *Yes* to accept (Figure 8).

Using the Connection File Method

Searching an online database using connection files consists of four basic steps.

A. Establishing a connection to the database from EndNote.
B. Searching the database.

FIGURE 8. Replacing old connection file warning

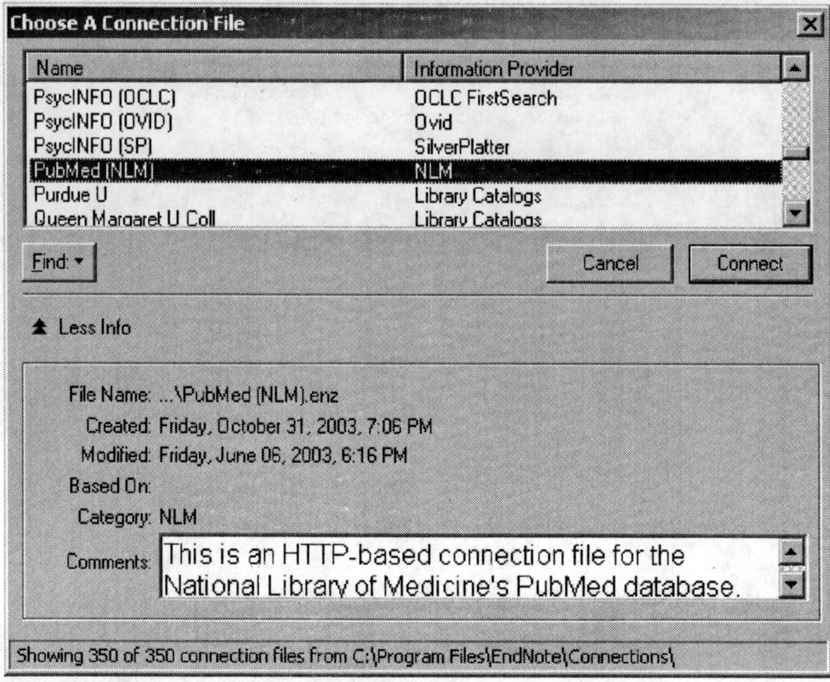

FIGURE 9. Selecting the PubMed connection file

C. Retrieving references.
D. Saving references into an EndNote library.

A. Establishing a connection to the database from EndNote

- Open an EndNote library or create a new library.
- Click *Tools* > *Connect* (Figure 5).
- Select *Connect* command again. Select from the list of databases in the next dialog box and click *Connect* (Figure 9). If you have connected to a particular database before, or if you have selected a database as "favorite," it will appear on the menu and you can click on it directly to connect.
- When a connection has been established, an empty Search window and an empty Retrieved References window (Figure 10) appear on the screen.

Technical Tip: The connection to the Internet database is maintained until you close the Retrieved References window or, after a period of inactivity, the connection automatically times out. If you are using a dial-up Internet connection (with a modem), EndNote does not disconnect you from the Internet after it closes the connection to the remote database.

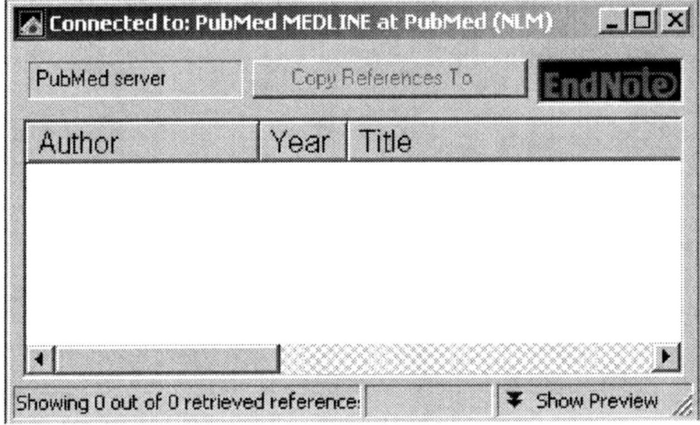

FIGURE 10. Empty Retrieved References window

B. Searching the database

Searching the remote database is essentially the same as searching your EndNote library. See Chapter 6 for a detailed discussion of the Search Window and search strategies:

- Enter your search terms in the Search window (Figure 11). Notice that the title bar of the search window displays the database you are connected to.
- Choose appropriate search options.
- Make sure "*Search Remote*" box is checked.
- Click *Search* button.

General Guidelines for Searching

- The options available in the search menus (such as Author, Title, or Keywords) vary with each database. In the next section of this chapter, you will learn more about searching PubMed.
- Click *Add Fields* to append search items to the list.
- Multiple search terms can be combined using the Boolean operators AND, OR, and NOT. Keep in mind that OR broadens your search by adding together two result sets; AND narrows the focus of the search by looking for the intersection of references found; and NOT also narrows the search, by omitting the results matching one term from the current results set (see Chapter 6 Figure 9).
- The comparison menu for each search term is always set to *Contains* for remote databases.

C. Retrieving references

EndNote displays the number of references found as a result of the Internet database search in a window (Figure 12). You can change one or both of these numbers to specify the range of references to be downloaded. Click *OK* to retrieve references in the Retrieved References window.

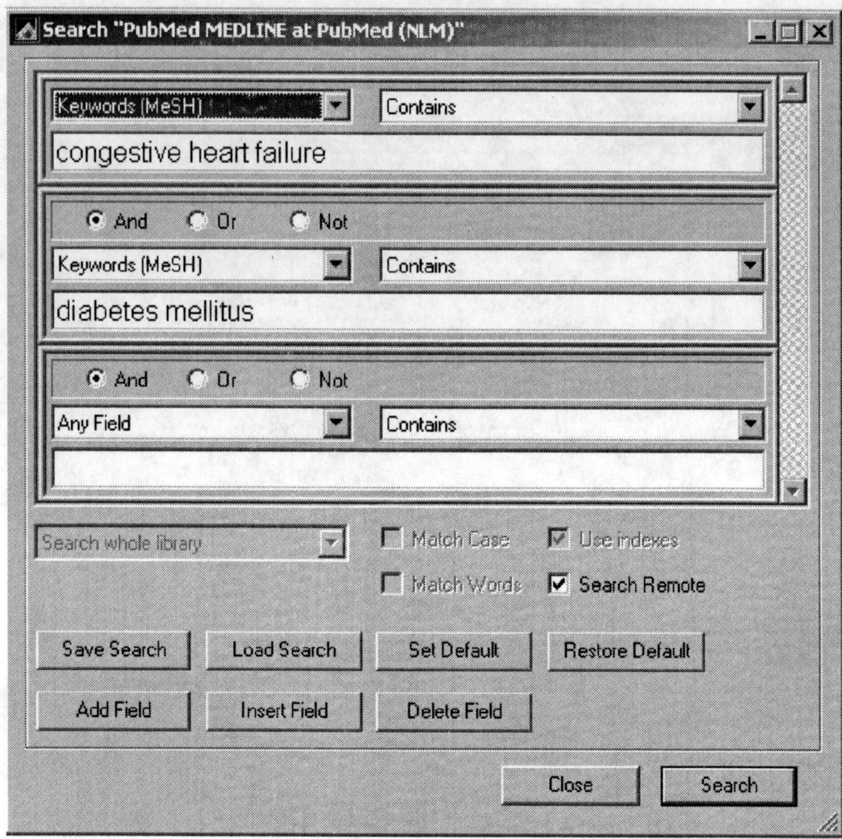

FIGURE 11. EndNote search window with search terms

Technical Tip: Be as specific as possible in your search strategy to avoid retrieving large number of references. A large number of references takes a long time to load as well and they will clutter your library.

Alert: The order of retrieved references reflects the way they were returned from the server—it is not necessarily alphabetical, chronological, or in order of relevance.

Working in the Retrieved Reference Window (Figure 13)

The window's **title bar** displays the name of the Internet database to which you are connected. The **message** area at the top of the window shows the progress of reference retrieval or the number of references currently displayed. The **status**

Confirm Remote Search

Found 877 References.

Retrieve References From: 1 through: 124

Discard the previously retrieved references?

OK Cancel

FIGURE 12. Window displaying the number of references found by the search

bar at the bottom displays messages pertaining to the status of the connection and reference retrieval.

- Double-click on a reference to view it.
- Click on the *Show Preview* button to preview a formatted version of the selected references in the preview pane.
- Click on the *Pause* button to pause or resume reference retrieval. Pressing the Escape (Esc) key also stops retrieval.

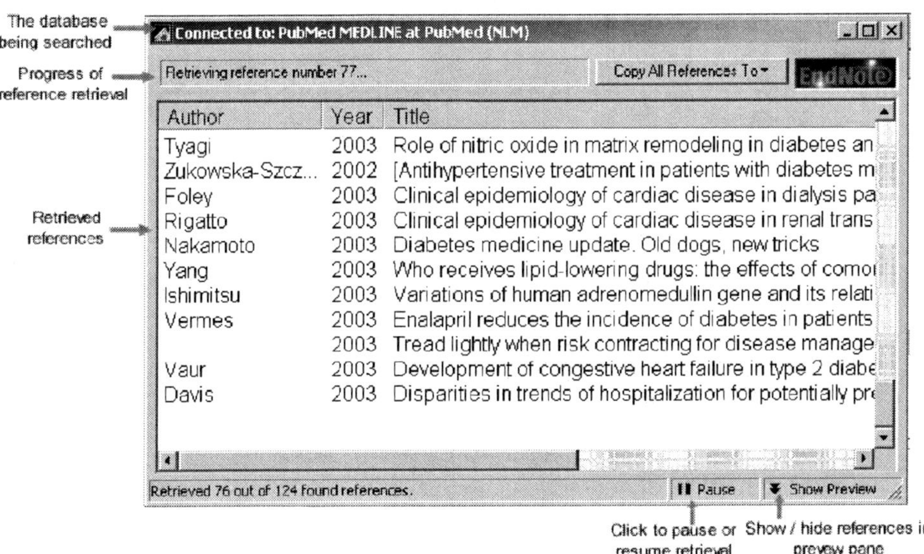

FIGURE 13. The Retrieved References window

- Use the *Search, Sort*, and *Show/Hide Selected References* commands to help you review the search results.

D. Saving references into EndNote library

You must now save references (selected few or all) into an EndNote library. The Retrieved References window is only a temporary holding place for references and can not store them permanently. When you close this window, all the retrieved references will be discarded.

Perform one of the following steps to save retrieved references into and EndNote library:

(a) If you want to save all references in this window to your library:
Do not select any references. Click on *Copy All References To* button and select a library to save the references.

(b) If you want to copy only selected references (clicking on a reference will highlight and select it):
- If the destination library is open, drag-and-drop selected references from the Retrieved References window into the EndNote library.
OR
- Click *Copy References To* button and select a library to save.
OR
- Click *Edit* > *Copy* menu, then paste into another library by opening the desired library and choosing *Paste* from the *Edit* menu.

> **Technical Tip:** EndNote does not check for duplicates when you use these methods. You should use the *Find Duplicates* command in the destination library to check for and delete duplicate references. See Chapter 6 for details.

The Import Filter Method

What is an Import Filter?

Filters allow you to import citations into your EndNote library by telling EndNote how to interpret the information you have saved from a particular database.

What Import Filters do I Have?

Before attempting to import records from a database into EndNote using a filter, check that an appropriate filter is available. The EndNote program includes over 350 filters. To see if a filter for a database is already available in EndNote:

- Click *Edit* > *Import Filters* > *Open Filter Manager* (Figure 14).
- The next screen displays a list of import filters available in EndNote (Figure 15). Notice that the features of this window are very similar to the connection manager window.

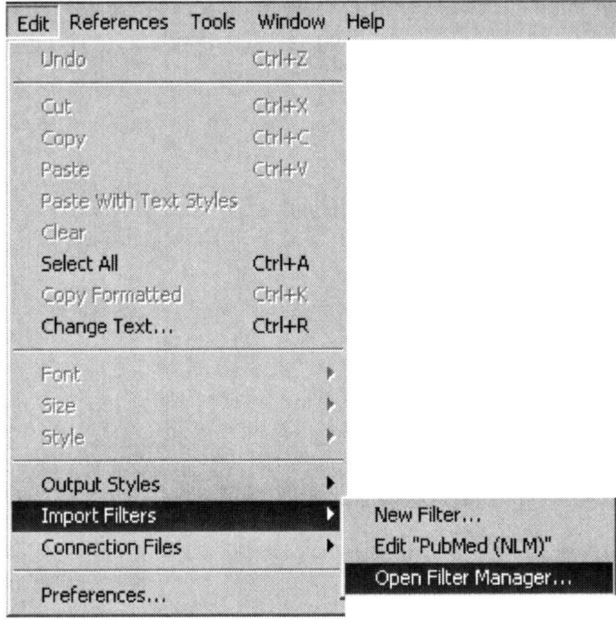

FIGURE 14. Opening the Filter Manager

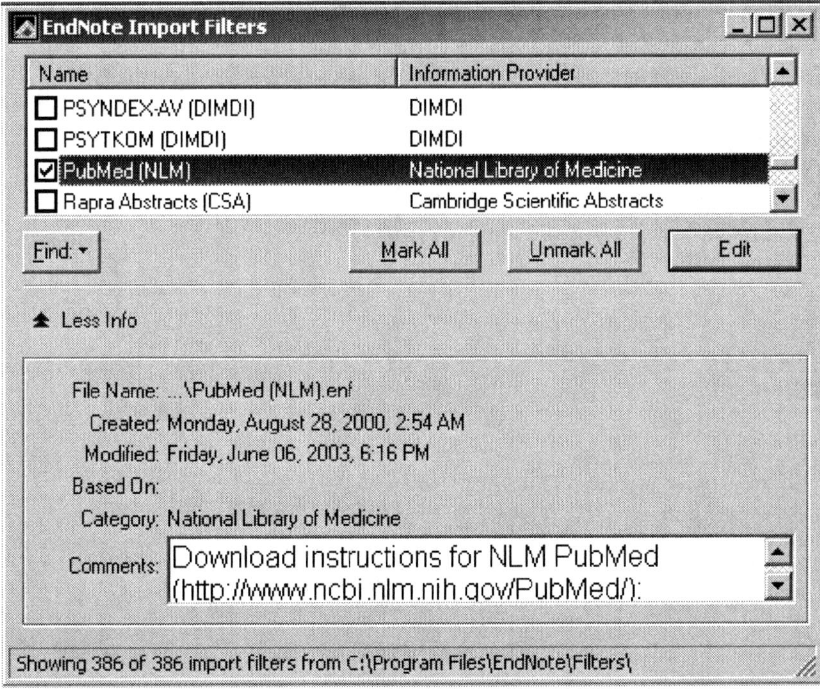

FIGURE 15. The Import Filter Manager window

If you can't find the filter of your choice, visit EndNote's web site at http://www.endnote.com/support/enfilters.asp. You can download the most up to date filters from this web site for a database of your interest.

Using the Import Filter Method

There are four basic steps involved in this method:

A. Searching the database, generally using a web browser.
B. Selecting references to be imported into EndNote.
C. Saving the selected references in a text file.
D. Importing the text file into an EndNote library using an **Import Filter**.

Steps demonstrating the use of the import filters depend on the database you are using because different databases offer different methods for saving the references in a text file. The import filter method can be best understood by looking at an example. See the section "Using EndNote with PubMed" to see a step-by-step example of this process.

The Direct Export Method

This method is essentially the same as the import filter method except that it automates the process of saving and importing references. Instead of using the intermediate step of the text file, you can do a direct export of selected references from the website of the Internet database into your EndNote library. EndNote uses the same import filters for direct export as for the import filter method. Some databases that offer a direct export feature include Web of Knowledge and ProQuest. For some other databases, you may have to download and install a filter before the direct export tool will work.

Table 3 lists some popular databases that allow direct export.

TABLE 3. Databases allowing direct export into EndNote

Database	Information provider	URL
CINAHL, Medline, AMED, BIOSIS, ERIC, Compendex, INSPEC, PsycINFO, SocioFile, Wilson Art Abstracts	OVID	www.ovid.com
Business Source Premier, Academic Search Elite, EconLit	EBSCOHost	www.ebscohost.com
ABI/Inform, ProQuest Computing, ProQuest Social Science Journals	ProQuest	www.proquest.com
ScienceDirect	Elsevier	www.sciencedirect.com
Current Contents	Thomson ISI	www.isinet.com
Web of Science	Thomson ISI	www.isinet.com

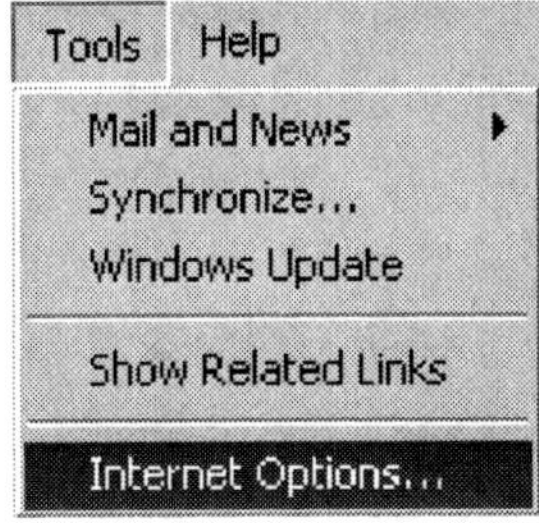

FIGURE 16. Selecting Tools > Internet Options in Internet Explorer

Technical Tip: Sometimes you will get an error message when trying to export citations from an online database into EndNote. The error message will display "File not found" and the records will not appear in EndNote. This happens because your cache of temporary Internet files is full and needs to be cleared. To do this:

- Click *Tools > Internet Options* in **Internet Explorer** (Figure 16).
- In the next screen, make sure the General tab is selected. Click on the *Delete Files* button under the *Temporary Internet Files* heading (Figure 17). Click *OK* in the next warning box.
- If you have a lot of temporary files, it may take a while to delete them. Once this is done, try to export the references again.

Similar to the import filter method, steps demonstrating the use of direct export depend on the database you are using and the direct export method can be best understood by looking at an example. See the sections "Using EndNote with Ovid" and "Using EndNote with the Web of Science" for step-by-step examples of this process.

Using EndNote with PubMed®

Introduction

PubMed is a service of the National Library of Medicine that includes over 14 million citations and author abstracts from more than 4,600 biomedical journals published in the United States and in 70 other countries, dating back to the 1950's. These citations are from the MEDLINE® database and some additional life science journals. Although coverage is worldwide, most records are derived from English language sources or have English abstracts. Abstracts are included for more than 75% of the records. PubMed also includes links to many sites providing full text articles and other related resources. You can access PubMed online at: http://www.ncbi.nlm.nih.gov/entrez/query.fcgi?db=PubMed.

MEDLINE is the National Library of Medicine's premier bibliographic database. The MEDLINE database is the electronic counterpart of *Index Medicus®*, *Index to Dental Literature*, and the *International Nursing Index*.

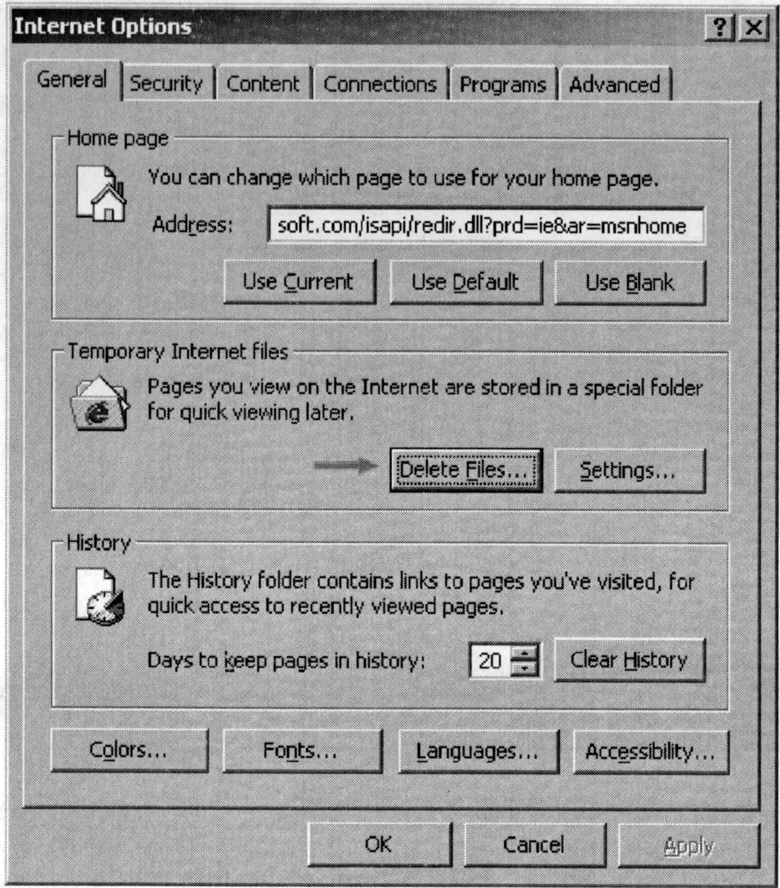

FIGURE 17. Deleting temporary Internet files in Internet Explorer

You can search and retrieve references from PubMed using the connection file method or the import filter method. The connection file method is the most direct and simple. However, if you would like to use PubMed's search features and thesauri to facilitate your search (which are certainly more extensive than EndNote's), then you should perform your search on PubMed and retrieve selected references into EndNote using the import filter.

Using the Connection File Method for PubMed

Remember: your computer must be connected to the Internet to use this feature.

- Open an existing EndNote library or create a new EndNote library.
- Click on *Tools* > *Connect*, and select *Connect (Figure 5)*.

- The next screen asks you to choose a connection file. Scroll down to select PubMed (NLM) where the information provider is NLM. Click on *Connect* to start the connection to PubMed (Figure 9). If you have connected to PubMed before or if you have selected PubMed as a "favorite," it will appear on the list and you can click on it directly (Figure 5).
- Depending upon the speed of your Internet connection, you will now be connected to the PubMed server.
- Next you will be presented with the search window. Enter your search parameters in the search window. Notice the Boolean search operators—And, Or, Not. Notice the format for author name. Click the Add Field button if you need more fields. Notice that you have only one choice in the comparison menu—"contains." Make sure the Search Remote box is checked. Figure 11 illustrates a sample PubMed search.
- After you are done inputting search parameters, click on the *Search* button.
- After a successful search, EndNote alerts you to the number of references that were found in a database (Figure 12).You have the option of retrieving all of the references or a specific range of references.

Click OK to retrieve references. In the next screen, copy all or selected reference(s) to the EndNote library.

Technical Tip: Remember, a single EndNote library can hold only 32,267 records. Once a record number is assigned to a reference, it can never be assigned again. So if you input 32,000 references into an EndNote library and delete all of these, you can still input only 267 more references into this library. Therefore, if you are going to be importing large number of records, I suggest that you create a temporary mini library for imported records, screen the records, and only keep the pertinent ones in your permanent EndNote library.

Creating a mini-library to import into master library later can also be used to eliminate duplicates and to apply keywords to the set before importing references into the main library. See Chapter 6 for how to eliminate duplicates and perform group editing of references.

Alert: Recently, the National Library of Medicine implemented many changes in their PubMed database architecture and infrastructure. This has caused the *PubMed (NLM)* connection to fail with older versions of EndNote. Currently, the following EndNote versions are compatible with connecting and searching PubMed:

- EndNote 7 (Windows and Macintosh).
- EndNote 6.0.2 (Windows and Macintosh)—Users of this version will need to download an update from the vendor's website (http://www.endnote.com/support/enupdates.asp) to restore the Connect feature in EndNote.

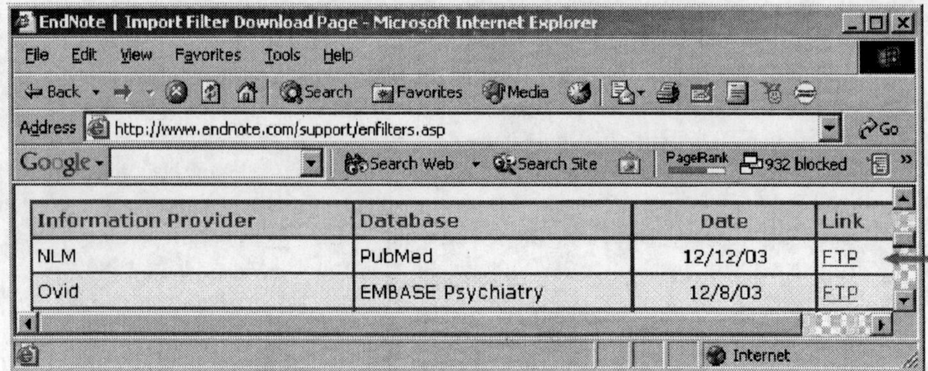

FIGURE 18. Selecting the NLM filter to download

Using the Import Filter Method for PubMed

Important note before you start

If you are using EndNote 7.0, note that you may not be able to successfully retrieve references into EndNote from the PubMed database due to recent architectural changes to the database. To fix this problem, you must download the new PubMed filter from EndNote's web site.

Steps to download the latest PubMed filter

1. Open the filter download page on EndNote's web site in your browser at http://www.endnote.com/support/enfilters.asp.
2. Locate the NLM filter from the list. Click on the "FTP" link for the filter (Figure 18).
3. Click *Save* in the next dialog box (Figure 19).
4. Save this file in the Filters folder within the EndNote folder. The most likely path for this file would be C:\Program Files\EndNote\Filters, but this path may vary depending on your installation. However, if you used the default options during EndNote installation, this is where you should save your EndNote filter file.

These steps can be followed to download any other filters you need from EndNote's web site.

Steps for Using the Import Filter Method for PubMed

- Open the PubMed website in a browser window at http://www.ncbi.nlm.nih.gov/entrez/query.fcgi?db=PubMed
- Perform a search in PubMed. I used the search term—"Computerized Patient Record System" AND "Medical Errors" (Figure 20).
- Select desired references from the list by checking boxes next to them (Figure 21).

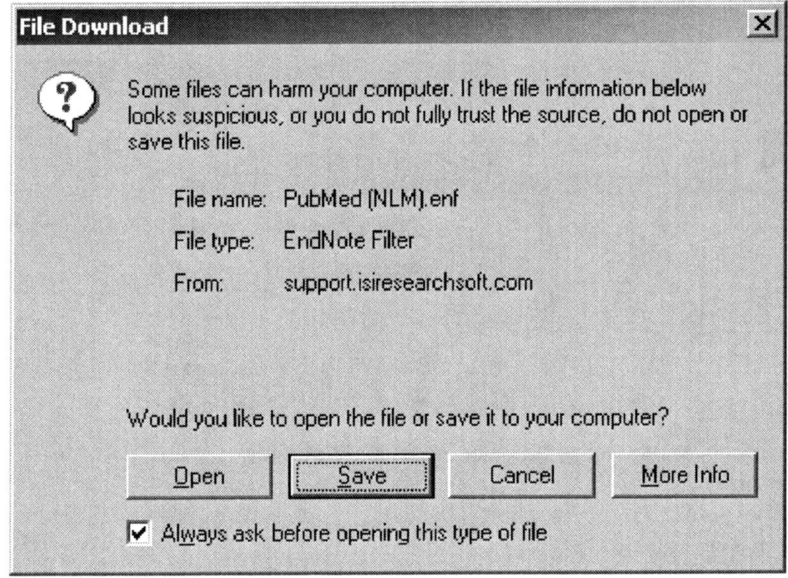

FIGURE 19. Saving the NLM filter

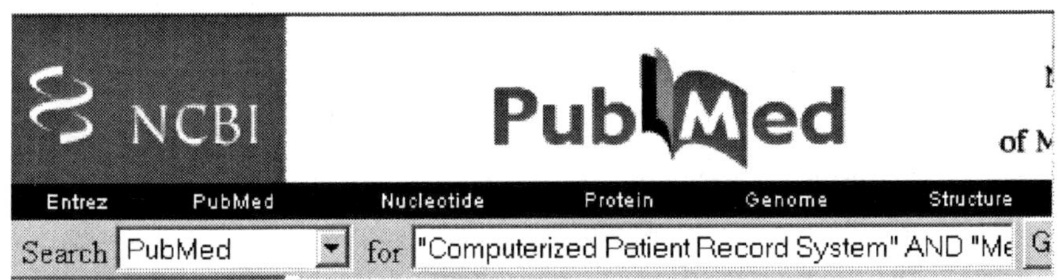

FIGURE 20. Performing a search on PubMed's web site

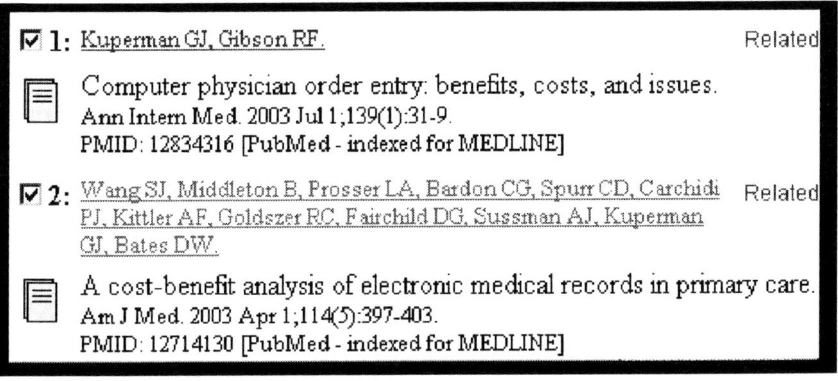

FIGURE 21. Selecting references from the PubMed search

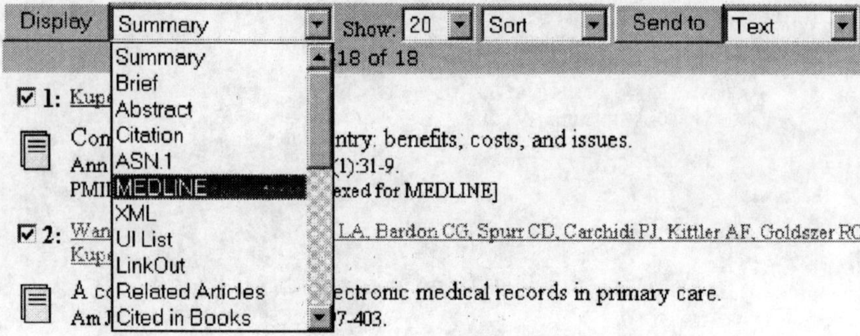

FIGURE 22. Selecting MEDLINE from the Display menu

- Scroll to the bottom of the window. From the Display menu, select MEDLINE (Figure 22).
- From the *Send to* menu, select File and then click the *Send to* button (Figure 23).
- In the next dialog box, click *Save* (Figure 24).
- Save this file using any name. Make sure you select *All Files* from the pull down menu and you type ".txt" at the end of the file name as EndNote can import only plain text files and not rich-text format (RTF), Word, or WordPerfect files. In this example, I named the file pubmed.txt and saved it on my *Desktop* (Figure 25).

> ⚠ **Alert:** You can save this file anywhere—on your hard drive, floppy drive, or a network drive—but it is **very important** that you remember the location where you saved the file. When you import this file into EndNote in the next step, you will need to specify the location of this file in computer. EndNote does not have a way of knowing the location of your file. If you saved the file on your desktop your file will look like this (Figure 26).

- Close PubMed—you no longer need it open and active.
- Start EndNote.
- In EndNote, open the EndNote library you want to import the data into.
- Click *File > Import* (Figure 27).

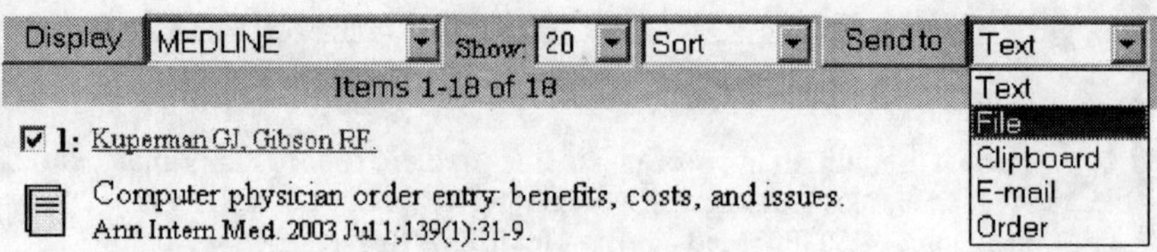

FIGURE 23. Selecting File from the Send to menu

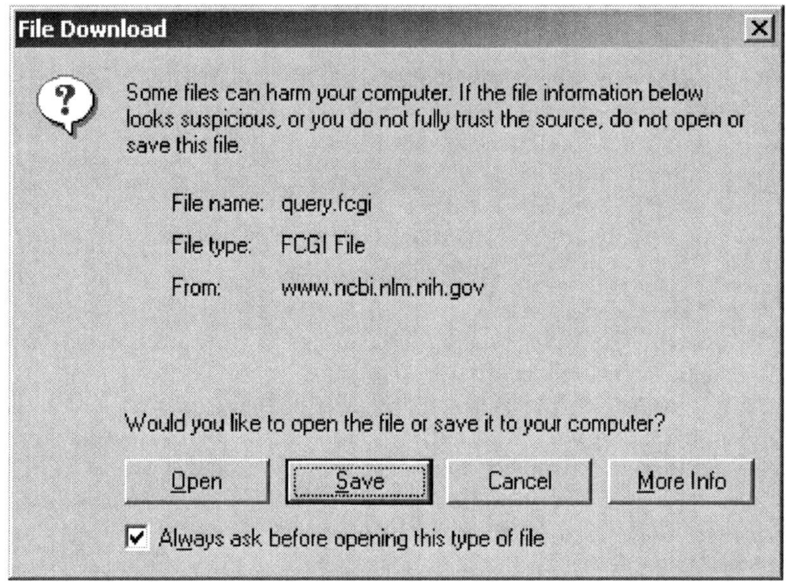

FIGURE 24. Click Save to download the PubMed search results file

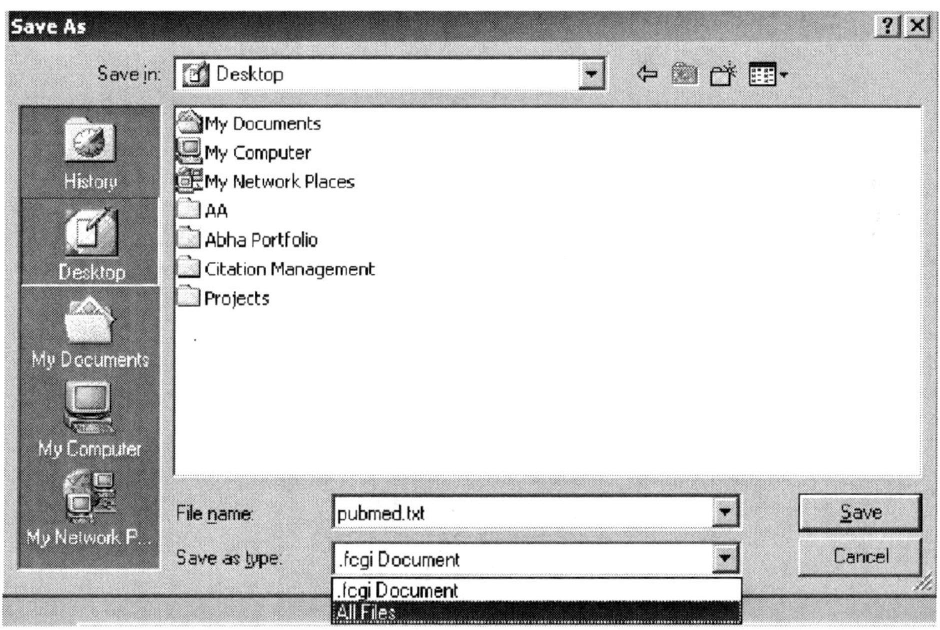

FIGURE 25. Selecting the file type and location for the PubMed search results file

FIGURE 26. PubMed text file icon on the desktop

- The next window presents the Import options (Figure 28).
  - Click *Choose File* button to locate and open the text file you saved from PubMed containing the references you want to import.
  - Click on the *Import Option* button to select the filter you will use. You **must** use the filter named "PubMed (NLM)" for the import process to work properly.
  - Select your preferred option from the *Duplicates* pull down menu. The Text Translation option can be used to improve EndNote's importing of accented characters. Change this setting *only* if you find that accented characters are not importing correctly.
- Click the *Import* button to start the import process. The citations will be imported into your library. **Tip: Only the citations you just imported will show in the library window** (Figure 29).
- To display all references in the library, click *References* > *Show All References* (Figure 30).

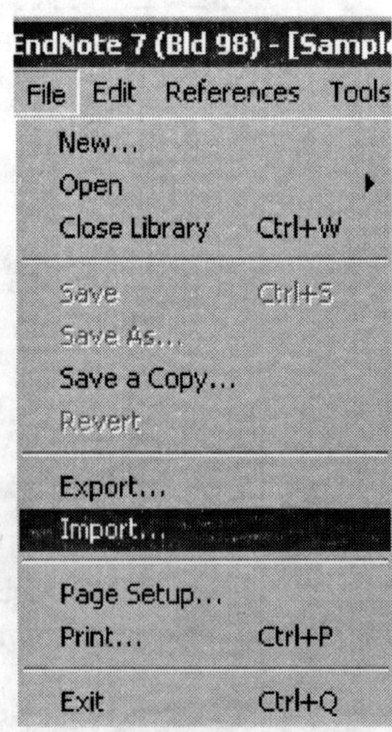

FIGURE 27. Selecting File > Import

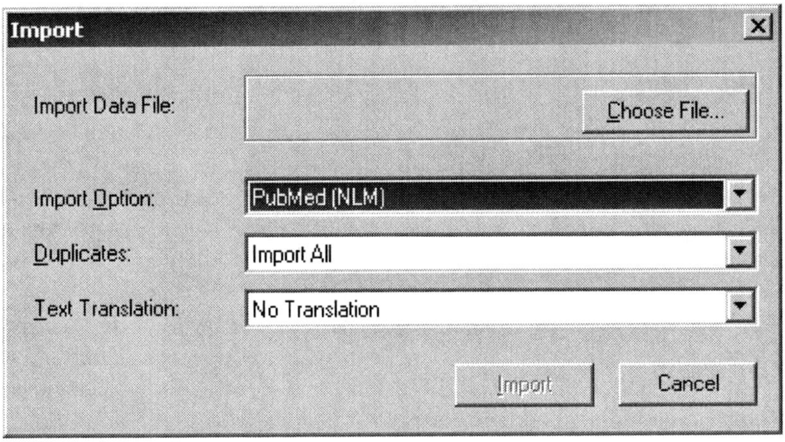

Import
Import Data File:
Choose File...
Import Option:
PubMed (NLM)
Duplicates:
Import All
Text Translation:
No Translation
Import
Cancel

FIGURE 28. Selecting import options

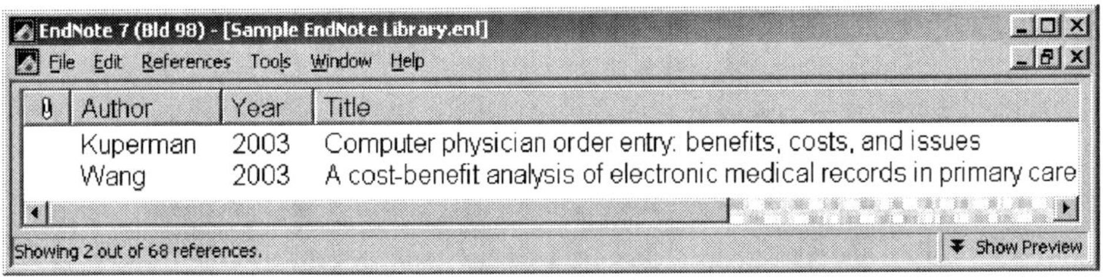

EndNote 7 (Bld 98) - [Sample EndNote Library.enl]
File Edit References Tools Window Help
Author
Year
Title
Kuperman 2003 Computer physician order entry: benefits, costs, and issues
Wang 2003 A cost-benefit analysis of electronic medical records in primary care
Showing 2 out of 68 references.
Show Preview

FIGURE 29. EndNote window showing only the imported references

EndNote 7 (Bld 98) - [Sample EndNote Library.enl]
File Edit References Tools Window Help
New Reference Ctrl+N
Edit References Ctrl+E
Delete References... Ctrl+D
Search References... Ctrl+F
Go To... Ctrl+J
Next Reference Ctrl+Page Down
Previous Reference Ctrl+Page Up
Show All References Ctrl+M
Show Selected References
Hide Selected References
Sort References...

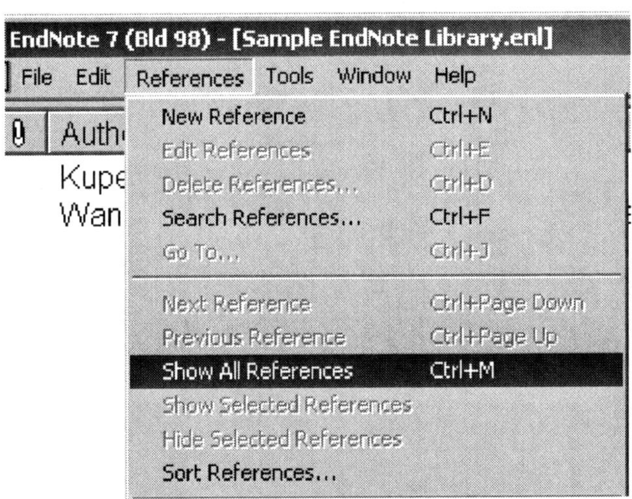

FIGURE 30. Choosing Show All References

Using EndNote with Ovid®

Introduction

Ovid is an online service that provides access to many biomedical databases including MEDLINE, ACP Journal Club, Cochrane database, Biosis, Embase, and PsycInfo, as well as tools for searching and navigating these databases. You can access the general information website for Ovid at http://www.ovid.com. Although PubMed is freely available to the public online, Ovid offers only subscription-based access.

EndNote allows you to retrieve references from the Ovid database using the connection file method as well as direct export. In my opinion, the direct export method is preferable for several reasons. First, to search Ovid with EndNote using the connection file method, you need to have an individual password instead of an institutional account. Second, by using direct export you can take advantage of Ovid's user-friendly and intuitive search interface. Third, the direct export feature eliminates the need to create an import file and selecting an import filter. This makes the direct export practically as simple and convenient as the connection file method.

Using the Direct Export Method for Ovid

- Perform a search on an Ovid database, such as MEDLINE (Figure 31).
- Select the references that you wish to retrieve into an EndNote library from the search results (Figure 32).
- Scroll down the web browser screen to **Citation Manager** and make the following selections (Figure 33):
  - Choose *Selected Citations* under Citations.
  - Select the fields for export to EndNote. Include Subject Headings for better indexing of your EndNote library in future. I have selected the "Citation + Abstract + Subject Headings" option.

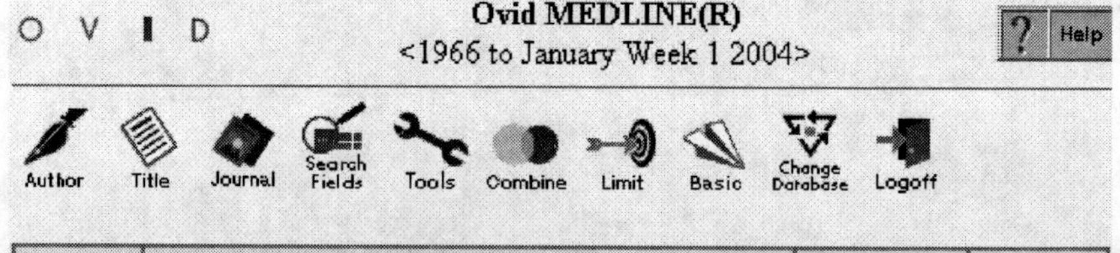

#	Search History	Results	Display
1	Heart Failure, Congestive/	42933	Display
2	Diabetes Mellitus/	59980	Display
3	1 and 2	575	Display

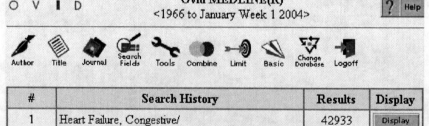

FIGURE 31. Performing a search in Ovid

☑ 1. Domanski M. Krause-Steinrauf H. Deedwania P. Follmann D. Ghali JK. Gilbert E. Haffner S. Katz R. Lindenfeld J. Lowes BD. Martin W. McGrew F. Bristow MR. BEST Investigators. The effect of **diabetes** on outcomes of patients with advanced **heart** failure in the BEST trial. [Journal Article] *Journal of the American College of Cardiology. 42(5):914-22, 2003 Sep 3.*
UI: 12957443

Abstract • Complete Reference

☑ 2. Bell DS. **Heart** failure: the frequent, forgotten, and often fatal complication of **diabetes**. [Review] [98 refs] [Journal Article. Review. Review Literature] ***Diabetes** Care. 26(8):2433-41, 2003 Aug.*
UI: 12882875

FIGURE 32. Selecting references from the Ovid search result

- Choose *Direct Export* under Citation Format.
- Click *Save*.

• The next screen will present an EndNote dialog box. Select the EndNote library into which you would like to export these references. Click on *Open*.
• The references will be imported automatically into the library.

Using EndNote with the Web of Science®

Introduction

ISI's Web of Science consists of five high-quality databases containing information gathered from thousands of scholarly journals in all areas of research. The databases include:

• Science Citation Index Expanded™.
• Social Sciences Citation Index®.
• Arts & Humanities Citation Index®.
• Index Chemicus®.
• Current Chemical Reactions®.

The three citation databases contain the references cited by the authors of the journal articles they cover. You can use these cited references as search terms. A cited reference search enables you to use a published work that you know, to find other, later works that cite it. In addition to cited reference searching, you can search these databases by topic, author, source title, and address.

Citation Manager: Display, Print, Save, or Email Citations

Citations	Fields	Citation Format	Action
◉ Selected Citations ○ All in this set (1-2) and/or **Range:** [ ]	○ Citation (Title,Author,Source) ○ Citation + Abstract ◉ Citation + Abstract + Subject Headings ○ Complete Reference Select Fields	○ Ovid ○ BRS/Tagged ○ Reprint/Medlars ○ Brief (Titles) Display ◉ Direct Export ☐ Include Search History	Display Print Preview Email Save

FIGURE 33. Selecting direct export options in the Ovid Citation Manager

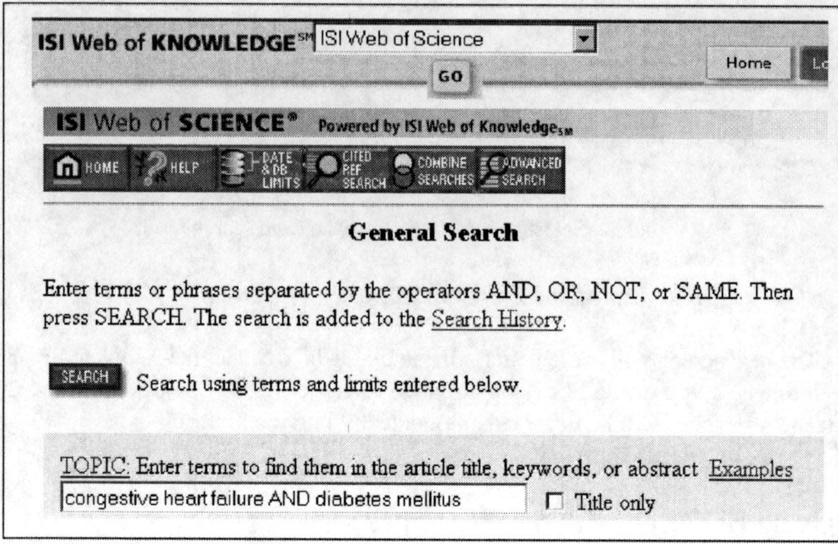

FIGURE 34. Performing a search in the Web of Science database

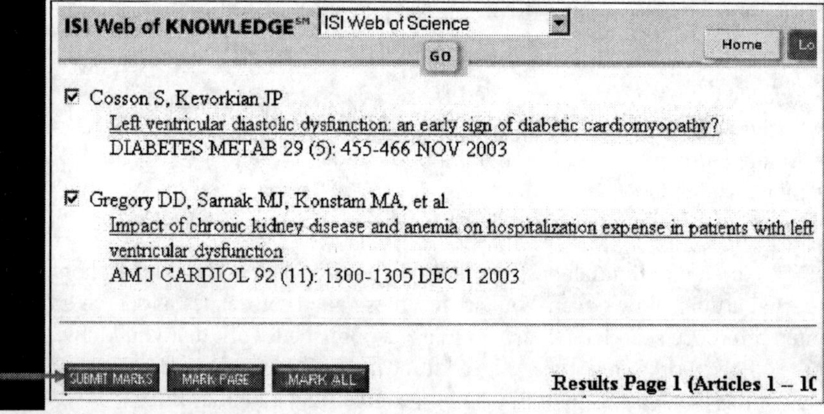

FIGURE 35. Selecting citations for direct export in the Web of Science

FIGURE 36. Clicking on Marked List to start direct export

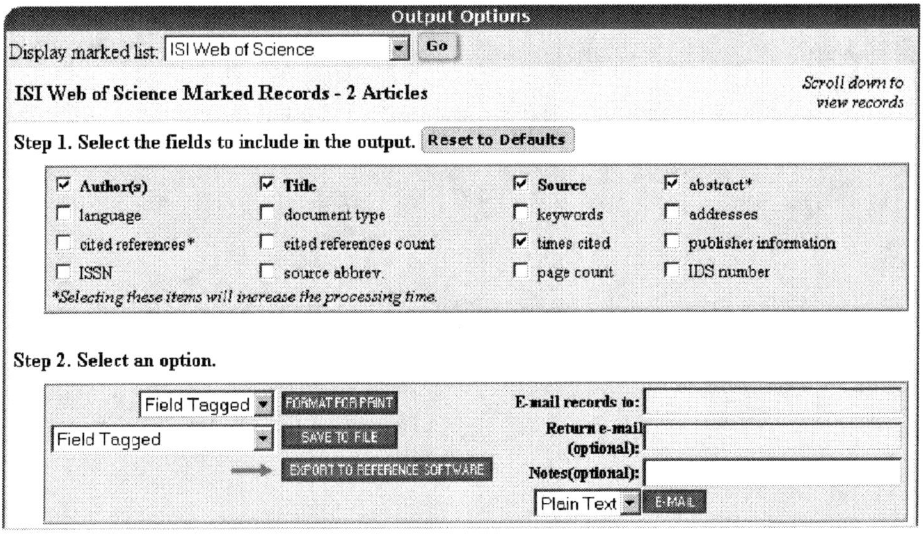

FIGURE 37. Selecting output options for direct export from the Web of Science

The two chemistry databases contain graphic representations of the structures and reactions reported in the journal articles they cover. You can search these databases by structure, by compound/reaction details, and by bibliographic information (topic, author, source title, and address).

Using the Direct Export Method for the Web of Science

As with Ovid, the direct export method is the easiest way to retrieve references from the ISI Web of Science into EndNote.

- Perform a search in Science Citation Index database in the Web of Science (Figure 34).
- Select references you wish to retrieve into EndNote by checking the box next to them. Click *Submit Marks* (Figure 35).
- Click on *Marked List* (Figure 36).
- The next screen presents you with "output options." Select additional fields to save in your library, such as 'keywords,' 'times cited,' and 'abstract.' Click on 'Export to Reference Software' to initiate the direct export process (Figure 37).
- The next screen will present an EndNote dialog box. Select the EndNote library where you would to export these references. Click on *Open*.

The citations will be imported into your EndNote library. Note that only the citations you just imported will show in the library window. To display all references in the library, click on the menu *Reference* > *Show All References*.

8
Creating Bibliographies Using EndNote

Things You will Learn in This Chapter

- A summary of steps in inserting citations and creating a bibliography.
- About output styles and working in the style manager to perform tasks such as marking a favorite style and editing or creating a new style.
- How to create a manuscript using EndNote manuscript templates.
- How to insert references from an EndNote library into a manuscript.
  - How to perform various operations on existing citations such as editing, deleting, moving, and unformatting.
- How to create bibliographies.
  - How to customize the font, layout and placement of a bibliography.
  - How to find and edit cited references in a library.
  - How to create a bibliography from multiple documents.
  - How to include notes in a list of references.
- How to work with Figure and Chart/Table reference types.
- About field codes, traveling library, and sharing your manuscript.
- How to cite references in footnotes.
- How to create independent bibliography, subject bibliography, and subject lists.
- How to set CWYW preferences.
- About miscellaneous tasks, such as turning off the field shading.

An Overview of Steps in Using EndNote to Create Bibliographies

Now that your EndNote library has references, let us discuss how to insert selected references from the library into a manuscript to create accurately formatted in-text citations and bibliographies.

EndNote provides a tool called Cite While You Write (CWYW), which helps you insert citations and format bibliographies in a manuscript. This tool is integrated with word processors, such as Microsoft Word, so that inserting references into a manuscript can be done without having to leave Word. Chapter 3 describes what to do if the CWYW commands are not properly installed into Word during EndNote installation.

Below is a summary of steps involved in inserting citations and creating bibliography in a manuscript using EndNote. Figure 1 shows the CWYW commands for these steps.

I. Create a manuscript in Word. You may use document templates provided by EndNote to create a paper for a specific journal such as Nature.

II. Insert EndNote references into manuscript
   - i. Place cursor in manuscript where you would like the in-text citation to appear.
   - ii. Click *Tools > EndNote > Insert Selected Citations.*

III. Format the bibliography
   - i. Click *Tools > EndNote > Format Bibliography.*
   - ii. Note that if you have "instant formatting" enabled, your bibliography will be formatted automatically after references are inserted. Click *Tools > EndNote > Format Bibliography* to select a different output style, font, or layout for bibliography.

IV. Send paper to a publisher
   - i. Remove special codes from the paper, by clicking *Tools > EndNote > Remove Field Codes*, to avoid incompatibility with publisher software.

The CD accompanying this book contains a sample EndNote library as well as a manuscript so you can practice various tasks discussed in this chapter before you have had a chance to create your own manuscript.

Output Styles

The bibliography creation process is critically dependent on the selection of an appropriate output style; therefore I have included this section in the beginning of the chapter. Understanding output styles and how to manage them will enable you to create bibliographies that meet your specifications.

The **output style** (also called just **style**) determines the selection of elements from various EndNote reference fields and sequencing, punctuation, and styling of references in bibliography as well as in-text citations (see Chapter 1 for an introduction to common referencing styles).

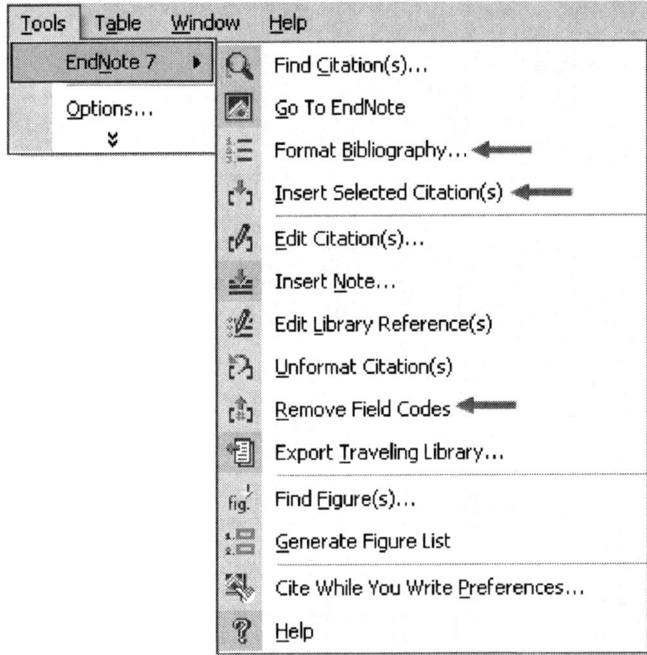

FIGURE 1. Frequently used CWYW commands in Microsoft Word

The choice of the output style will depend upon the publication to which you intend to submit your manuscript. Many biomedical journals have 'authors' guidelines,' including specifications for bibliography styles, available on the Internet. Table 1 lists some premier biomedical journals and the URLs for looking up their authors' guidelines.

TABLE 1. Authors' guidelines for some biomedical journals

Journal	Publisher	URL for authors' guidelines
The New England Journal of Medicine	Massachusetts Medical Society	http://authors.nejm.org/Misc/NewMS.asp
Annual Review of Immunology	Annual Review	http://www.annualreviews.org/authors/authors.asp#instrux
Annals of Internal Medicine	American College of Physicians	http://www.annals.org/shared/author_info.shtml
Science	American Association for the Advancement of Science	http://www.sciencemag.org/feature/contribinfo/home.shtml
Nature	Nature Publishing Group	http://npg.nature.com/npg/servlet/Content?data=xml/05_sub.xml&style=xml/05_sub.xsl
British Medical Journal	BMJ Publishing Group	http://bmj.bmjjournals.com/misc/ benchpress2.shtml

The Style Manager

EndNote provides over 1,000 predefined output styles, such as Vancouver, APA 5th, Life Sciences, and styles specific to many biomedical journals. These styles are stored as individual files in the Styles folder of the EndNote folder. You can use the **Style Manager** to view available styles, create a new output style of your choice, and customize any of the existing output styles to your specification.

You can reformat your in-text citations and bibliography at any time using a different output style just with a mouse click. This is particularly useful if you are going to resubmit your manuscript to a publication that has a different style requirement.

Working in the Style Manager

- To access the style manager: click *Edit > Output Styles > Open Style Manager* (Figure 2).

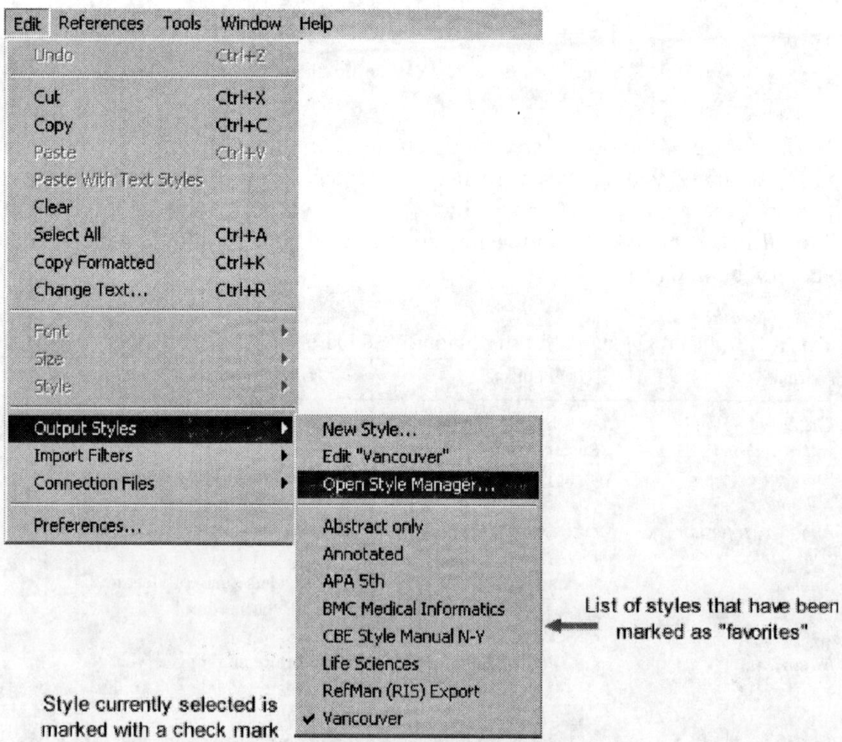

FIGURE 2. The Output Styles menu in EndNote

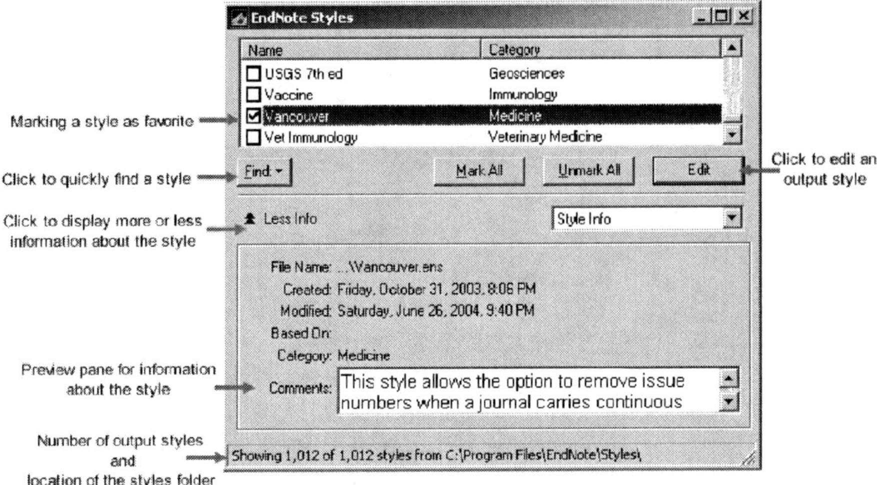

FIGURE 3. The Output style manager

The Style Manager Window

Figure 3 shows various features and commands available in the style manager window.

Marking Styles as Favorites

If there are some styles that you use frequently, you can mark them as 'favorites.' All styles marked as 'favorites' appear in the output styles menu so that you can easily access them without opening the style manager (Figure 2).

To mark a style as favorite,

- Check the box next to the style in style manager (Figure 3).
- Use the *Unmark All* button to unselect all previously selected styles.

Favorite styles also appear in the styles list in Word when you format bibliography by clicking *Tools > EndNote > Format Bibliography* **in Word** (Figure 4).

Editing Styles

You can modify selected components of an output style. To do this:

- Click *Edit > Output Styles > Open Style Manager.*
- Select a style by highlighting it.
- Click *Edit.*
- The next Style Editor window allows you to edit any style to your specification (Figure 5). You will see options to specify Anonymous Works, Page Numbers, Journal Names, Citations, Bibliography, Footnotes, and Figures and Tables.

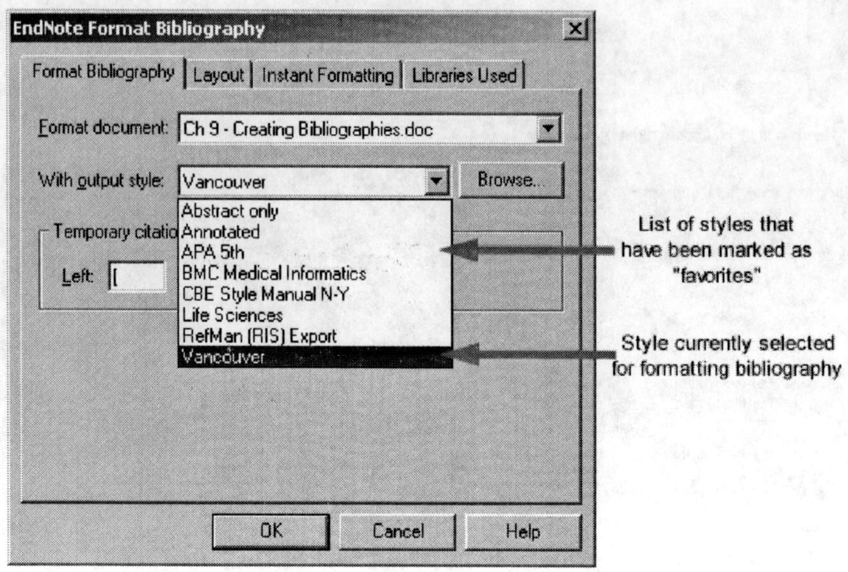

FIGURE 4. The Format Bibliography command in Word

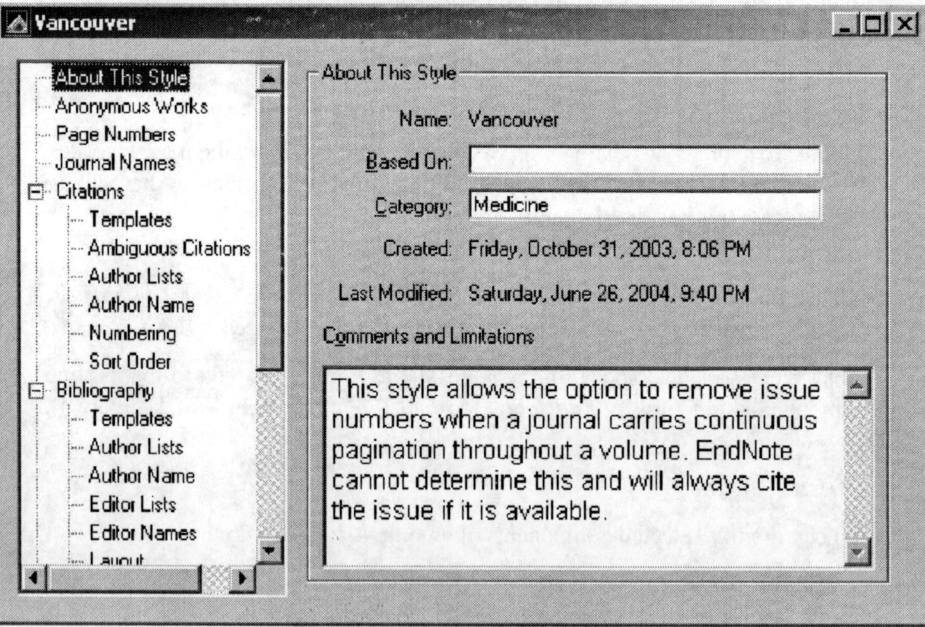

FIGURE 5. The Style Editor window

Note that although editing an output style will help you customize your bibliography precisely to your requirements, tinkering with a style may also have unintended consequences. My advice is that you use this feature judiciously.

Alert: I always make a copy of the output style I am modifying. This way, I have the original output style available to me, in case, a mistake is made during the editing or if I need to go back to the default style. To do this:

- Open Windows Explorer by clicking *Start* > *Programs* > *Accessories* > *Windows Explorer*.
- Navigate to the Styles folder; if you accepted defaults during the installation, the location for this folder should be C:\Program Files\EndNote\Styles.
- Right click on the style you would like to copy. Select *Copy* from the menu. Click *Paste* in the folder. You should see a copy of the style available in the Style Manager now.

Examples of Editing Styles

Chapter 9 includes an example of modifying the Vancouver style to include elements of an electronic reference such as URL, access date, and access year. Below is another example.

Excluding Author Names in the In-text Citation in the APA 5th Style

By default, the in-text citations in the APA 5th style are formatted as "(Author, Year)." Sometimes, this would create a problem as it may cause duplication of author names, for example, "In a report by Smith (Smith, 2001) the observation was...." There are two ways to address this problem:

Method 1: If you would like author name to be excluded from **all** in-text citations, the best approach is to modify the APA 5th output style to exclude "Author" names. To do this:

- Open the APA 5th style for editing by clicking *Edit* > *Output Styles* > *Open Style Manager*. Select APA 5th by highlighting it. Click *Edit*.
- Click on the Templates under the Citations heading. Delete "Author" and the punctuation immediately after it. Make sure the parentheses are intact (Figure 6).
- Reformat the bibliography using the APA 5th style. The in-text citation appearance in the paper should change from "(Smith, 2001)" to "(2001)."

Method 2: If you would like to exclude author names only from certain in-text citations and not all, then the best approach is to edit those individual citations to exclude "Author" names. To do this:

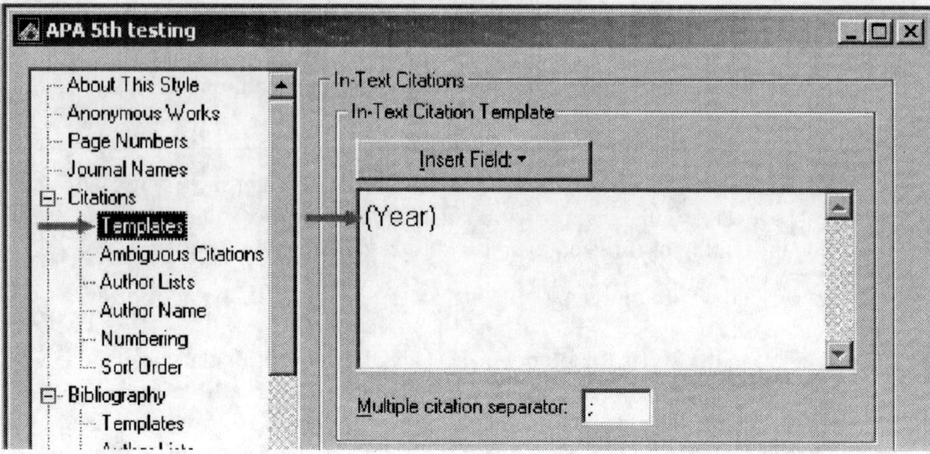

FIGURE 6. Editing the APA 5th Style to exclude Author names from the in-text citations

- Highlight the citation you want to edit. Right click.
- Select *Edit Citation* from the menu.
- Click on the check box next to the "Exclude Author Name" in the next screen to make sure it is selected (Figure 7).
- Fill in the information in the EndNote Manuscript Wizard.

FIGURE 7. The Edit Citations dialog

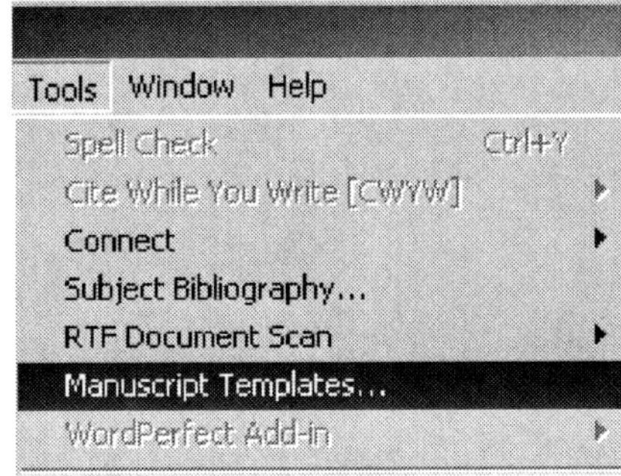

FIGURE 8. Clicking Tools > Manuscript Templates

Creating a Manuscript

You can create a manuscript using any word processor. EndNote facilitates the process by providing a variety of document templates for journals (e.g., for Nature, Nature Medicine, etc.) that you can use with Microsoft Word to help you create manuscripts in the formats required by publishers. When you use one of these templates to start your paper, many formatting issues are already set up for your target publication, such as proper margins, headings, pagination, line spacing, title page, abstract page, graphics placement, and font type and size.

- Click *Tools > Manuscript Templates* **in EndNote** (Figure 8).
- In the next dialog box, select the template you wish to use. The example in Figure 9 shows selecting the template for Nature. Click *Open*.

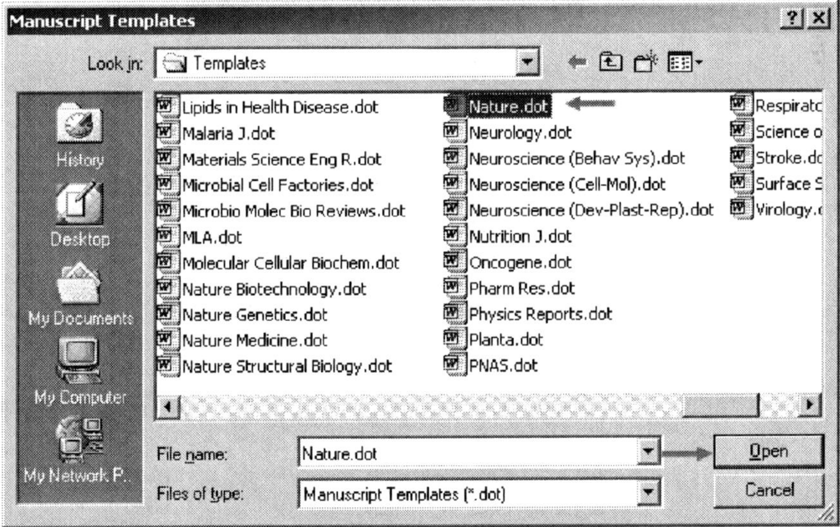

FIGURE 9. Selecting a manuscript template

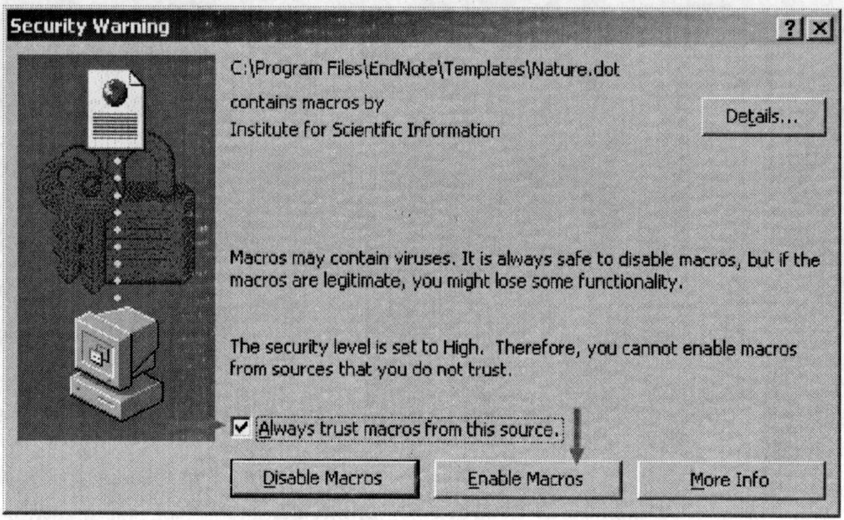

FIGURE 10. Enabling macros for manuscript templates

- If you get a security warning, check the box to trust macros from this source and click *Enable Macros* (Figure 10).
- This opens a Microsoft Word document in a new window based on the template file.

Inserting References from an EndNote Library into a Manuscript

Inserting References into Manuscript

- Open the EndNote library that contains the references you wish to cite. Open your manuscript.
- Position the cursor in the text of the manuscript where you would like to place the in-text citation.
- Click *Tools > EndNote 7 > Find Citation(s)* (Figure 11).
- The next screen shows the EndNote Find Citation dialog box (Figure 12). Note that if you have already selected and highlighted citation(s) in EndNote library that you would like to insert, you can simply click *Tools > EndNote 7 > Insert selected citation(s).*
- In the *Find Citation(s)* dialog box, write a search term in the *Find* box, and click *Search.*
- In the search results, select the citation(s) you would like to insert and click *Insert.*
- Insert as many citations as you need in the manuscript.

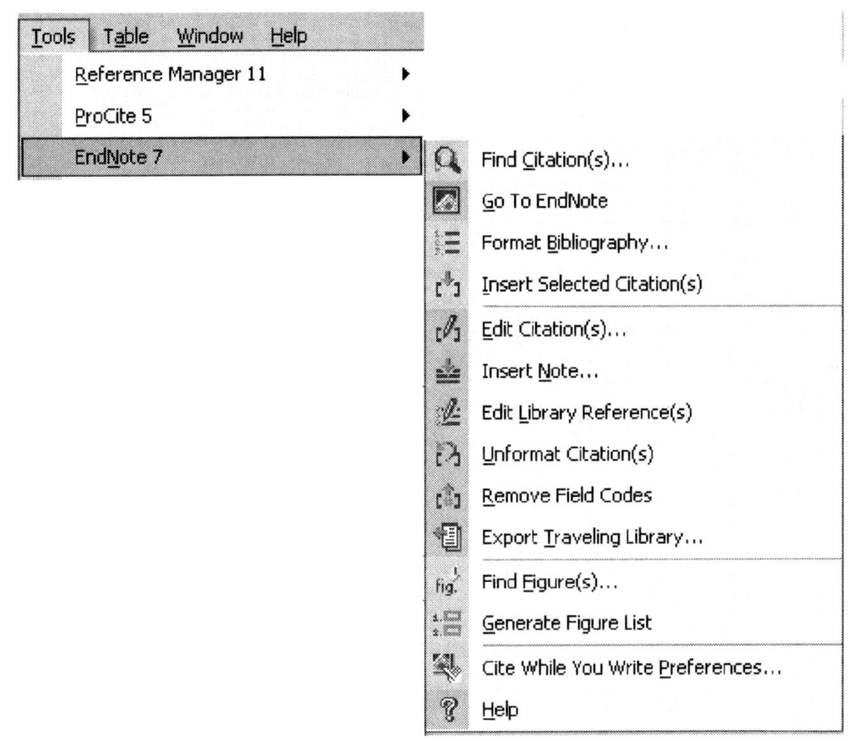

FIGURE 11. The CWYW menu

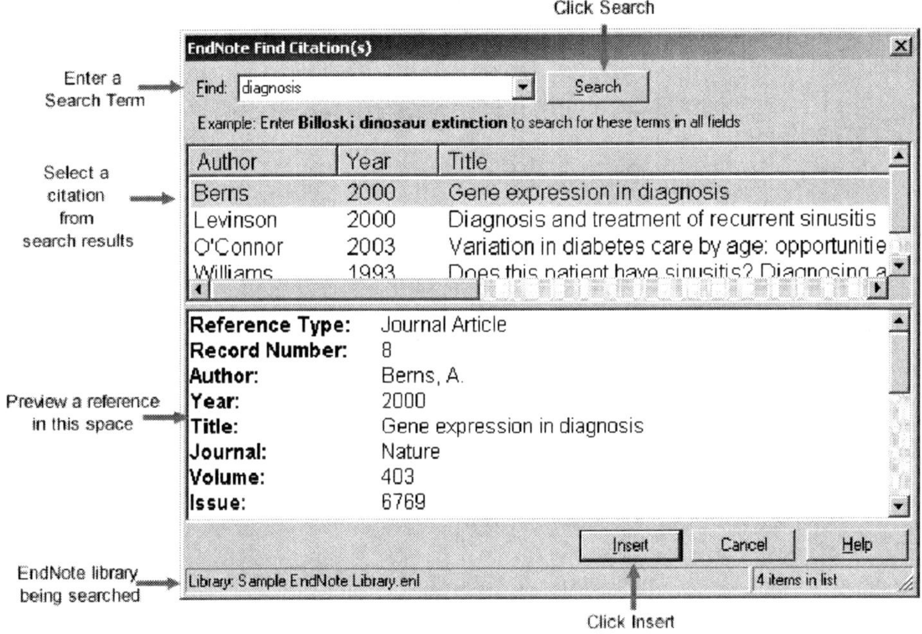

FIGURE 12. The Find Citations dialog

Changing Existing Citations

Editing Citations

This command edits the appearance of the in-text citation only and not the reference list/bibliography. To edit citations:

- Click on the citation you wish to change. **Note that your citation(s) must be formatted before you can use the Edit Citation(s) command.**
- Click *Tools* > *EndNote 7* > *Edit Citation(s).*
- In the next dialog box, select the citation you wish to edit (Figure 7). You have the option to make the following changes in the citation:
  - Exclude Author: to omit the author name from the formatted citation. Note that this option is useful only if you are using an output style that includes author names in citations.
  - Exclude Year: to omit date from the formatted citation. This option is also useful only if you are using an output style that includes dates in citations.
  - Prefix: enter text here to print immediately before the citation text.
  - Suffix: enter text here to print after the citation text.
  - Pages: enter page numbers here to print as "Cited Pages".
  - You can add or remove citations from a multiple citation, or change the order of citations by using *Insert, Remove*, or *Up* and *Down* arrow keys.
- Click *OK*.

> ⚠ **Alert:** Once you have inserted and formatted a citation, you should avoid changing it directly in Word because direct edits are lost the next time EndNote formats the bibliography. Instead you should use the *Edit Citation(s)* command if you must modify your formatted citations.

> **Technical Tip:** Sometimes the "Exclude Author" command does not remove the author name from citations. To fix this, you must ensure that EndNote preferences are set correctly:
>
> - Click *Edit* > *Preferences*.
> - Select *Formatting* heading (Figure 13).
> - Make sure "Omit author . . . " check box is checked.

Unformatting Citations

Unformatting reverts formatted citations to temporary citations. This feature allows you to easily identify references as you work in your document. A formatted citation may appear simply as a number in the text such as "(2)" while an unformatted citation appears as "{Williams, 1993, #21}" giving you an easy identification of the author's name, year, and the record number for this reference in EndNote.

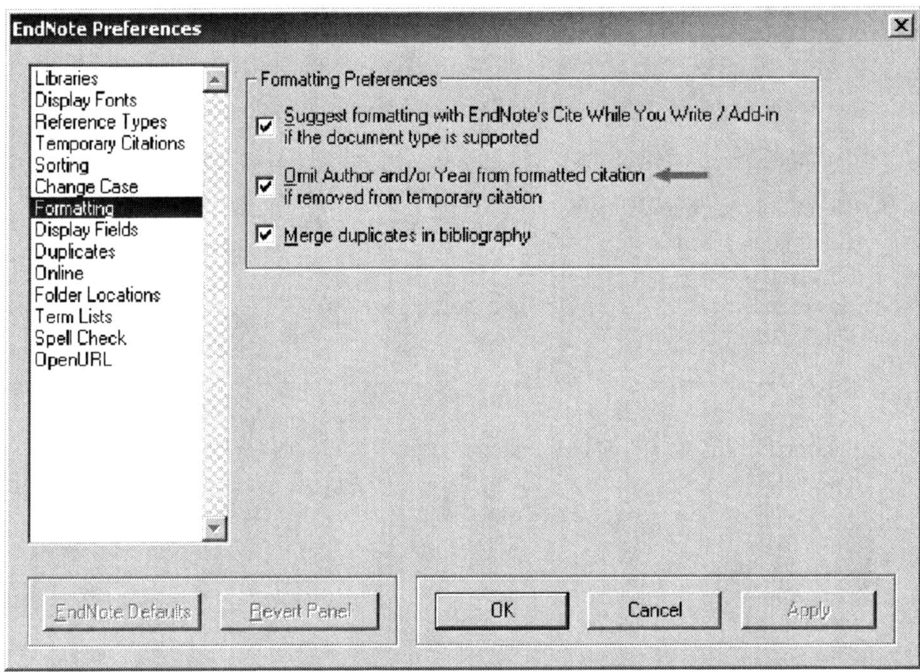

FIGURE 13. Setting EndNote formatting preferences

> ⚠ **Alert:** Unformatting removes the Traveling Library; therefore, you must have the corresponding EndNote library open in order to reformat the unformatted citation(s).
>
> However, unlike the Remove Field Codes command, unformatting does preserve the link between the paper and the EndNote library, allowing you to format citations again. See later sections in the chapter for details about traveling library and removing fields codes.

To unformat citations:

- Select the citation(s) you would like to unformat.
- In Word, click *Tools > EndNote > Unformat Citation(s)* (Figure 11).
- To unformat the entire document, either select nothing or highlight the entire document.

Moving or Copying Citations

You can move or copy formatted or unformatted citations using Word's *Edit > Cut* (or) *Copy > Paste* command. Make sure you highlight the entire citation, including surrounding delimiters.

Deleting Citations

To delete unformatted citations:

Simply highlight the citation and press the backspace or delete key.

To delete formatted citations:

- Highlight the citation you wish to delete.
- Click *Tools > EndNote 7 > Edit Citation(s)* (Figure 11). This opens up the Edit Citation(s) dialog box (Figure 7).
- Select the citation in the left column and click *Remove.*
- Click *OK*.

> ⚠ **Alert:** You should not delete formatted citations directly by pressing the 'delete' key in Word, because, if you do not completely delete the citation and all associated code (sometimes not easily visible), you could corrupt the document creating inaccuracies in the reference numbers and/or bibliography. Instead use the *Edit Citation* command.

Creating Bibliographies

Once you are done inserting all desired references from the EndNote library in the manuscript, you would need to format the in-text citation and a bibliography.

Formatting the Bibliography

If "Instant Formatting" is enabled (See "Setting Preferences for CWYW" section)

Your citations will be automatically formatted in the output style that was selected. You will begin to see in-text citations, as well as a bibliography in the paper, as soon as you finish inserting a reference from EndNote library into the manuscript.

If "Instant Formatting" is not enabled

- Click *Tools > EndNote > Format Bibliography (Figure 11).*
- A Format Bibliography dialog appears in which you can change or simply verify the output style and layout of the bibliography (Figure 4). Click *OK*.

Your paper now has sequentially numbered in-text citations and a reference list at the end of the paper. If you make any changes, such as adding or deleting a citation, simply click *Format Bibliography* again to update your bibliography.

Customizing the Bibliography

The Format Bibliography dialog box allows you to easily modify the layout, font, and so on, of a bibliography during formatting (Figure 4). This dialog box is

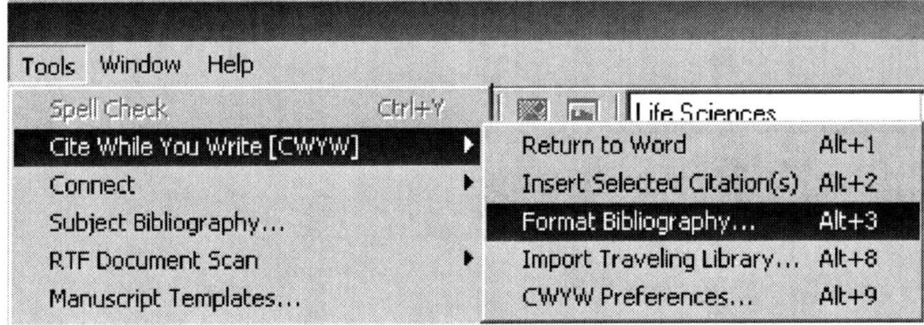

FIGURE 14. Accessing the Format Bibliography command from EndNote

accessed by clicking *Tools > EndNote > Format Bibliography* in Word or *Tools > CWYW > Format Bibliography* in EndNote (Figure 14).

Customizing the Font and the Layout

- Select the 'Layout' tab in the Format Bibliography dialog box (Figure 15).
- In this screen, you can change the following:
  - Font type and size of the reference list.
  - Bibliography title and text format.
  - Line spacing and indentation options for the reference list.

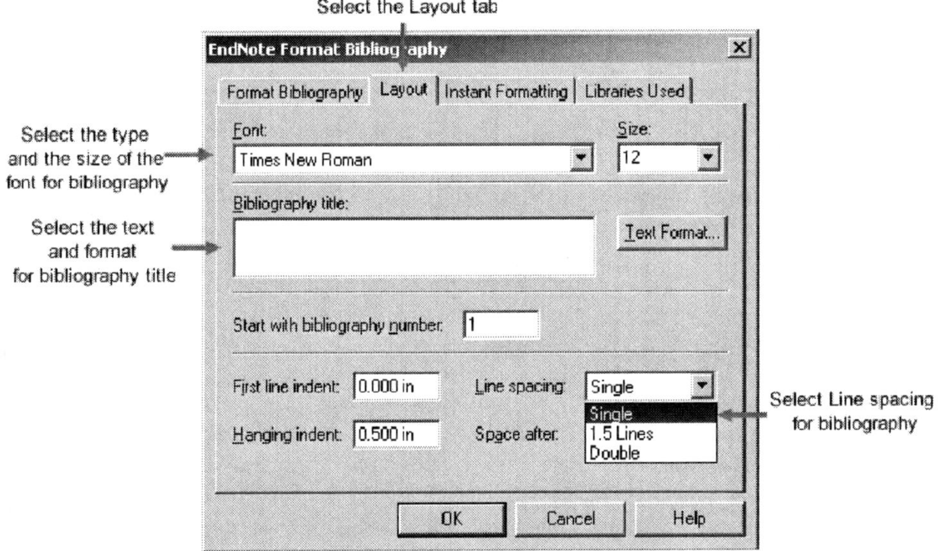

FIGURE 15. Customizing the font and layout of the bibliography

It is much better to make these changes in the Format Bibliography dialog box than to change the bibliography in your paper directly. Changes here will apply consistently every time you format the bibliography.

> **Alert:** This command does not change the font and layout in the captions of the Figure and Chart/Table type references.

Customizing Bibliography Placement

You may need to place a bibliography in the middle of a document, as opposed to EndNote's default placement of the bibliography at the end of the document. Using EndNote 7 and Word 2002, I have been able to perform this function without creating any custom styles or making any changes. Simply select the entire reference list/bibliography, then cut and paste it to the desired section of the Word document. When you add references or make changes to the bibliography, use the *Format Bibliography* command, and the bibliography stays there.

Finding and Editing Cited References in a Library

You may want to look up and edit the original EndNote reference linked to the in-text citation in a document. This is very simple to do:

- Make sure that the EndNote library, which contains references in the document, is open.
- Highlight the citation in the document.
- Click *Tools > EndNote 7 > Edit Library Reference(s)* (Figure 11)

EndNote opens the corresponding reference for editing. Note this command does NOT work for Figure and Chart/Table reference types.

> **Technical Tip:** If you make changes to an image or a chart/table in an EndNote reference, this change will NOT automatically be reflected in your Word document, even after repeating the *Format Bibliography* or the *Generate Figure List* command. I find the best way to achieve this is to delete the previous citation and perform an "insert citation" again.

Creating a Bibliography from Multiple Documents

This is particularly useful in situations such as creating a collective list of all references from multiple chapters in a book. You can use the "**Master Document**" feature of Microsoft Word to create a bibliography from multiple independent documents.

This procedure works the same way as formatting any bibliography; the trick is to create a master document in Word. Once you successfully create a master

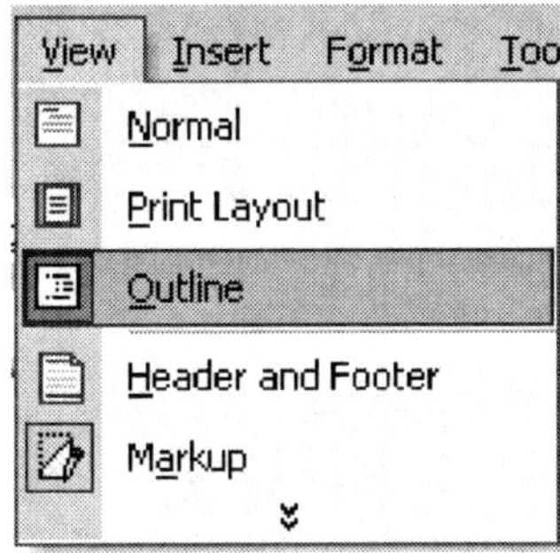

FIGURE 16. Selecting Outline View in Microsoft Word

document, along with desired chapters as subdocuments, you can simply call the usual CWYW's *Format Bibliography* command to create a cumulative bibliography.

To perform this:

- First, unformat citations in all the documents that contain citations/reference lists linked to an EndNote library. Click *Tools > EndNote > Unformat Citation(s)* command to do this (Figure 11).
- Create a new Word document, which will be the master document with the cumulative bibliography.
- Designate this new document as the Master Document in Word. To do this:
  - Click *View > Outline* in Word menu (Figure 16). You will notice that the appearance of the document has changed and it now shows a collapsible outline.
  - Create subdocuments by typing in the outline view headings (Figure 17).

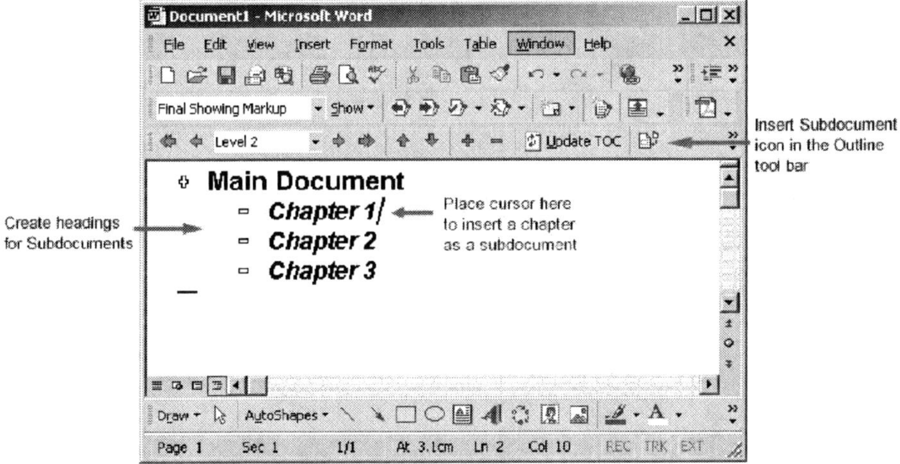

FIGURE 17. Inserting subdocuments in a master document in Word

- Insert chapters (documents with EndNote linked references) by clicking "Insert Subdocument" icon in the outline toolbar. Now you should see your chapters, along with their references, in this master document. *Hint: collapse the outline if you need.*
- Now that you have inserted each subdocument in this master document, click *Tools > EndNote 7 > Format Bibliography.*

> **Technical Tip:** Instead of creating a Master Document, I tried to use Word's *Insert > File* command to insert a chapter in the main document. This does NOT work. When I format the bibliography, EndNote formats only the citations in the main document and not in the linked document. Therefore, you do need to use the Master Document method to create a cumulative bibliography.

> **Alert:** There are multiple reports that, while useful, Master Documents are highly corruptible. Therefore, it is suggested that you create the master document only at the last step after finishing the editing of the original document. Secondly, always create copies of the original documents before 'merging' them in the master document.

Including Notes in the List of References

Some publications, notably the journal *Science*, require that you also include notes along with references in the reference list. These notes should have an in-text citation, should be sequentially numbered like references, and they should be included in the reference list. So your reference list might look something like this:

1. Reference 1
2. Reference 2
3. Note 1
4. Reference 3
5. Note 2

To insert a note:

- Place the cursor in the text of the document where you would like the in-text citation for the note to appear.
- Click *Tools > EndNote 7 > Insert Note* (Figure 11).
- EndNote Insert Note dialog window appears. Type in the text for the note. Click *OK* (Figure 18).

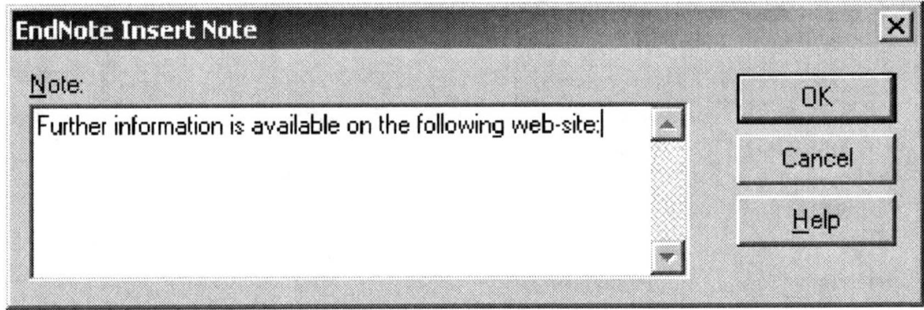

FIGURE 18. EndNote Insert Note dialog

- Click *Tools > EndNote 7 > Format Bibliography*. Now you should see a numbered citation for the note in the text of the document and the note included in the reference list.

Working with Figure and Table/Chart References in Manuscripts

Introduction

EndNote can insert and sequentially number figures and charts/tables in a manuscript by creating:

(a) In-text citations which appear as "(Figure 1)" or "(Table1)".
(b) A list of figures and tables (including actual image or table).

Figure 19 shows the basic components of a figure type citation. The same structure applies to tables and charts.

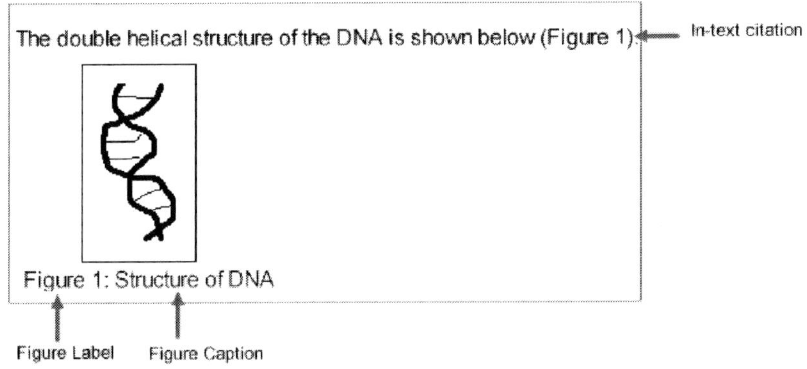

FIGURE 19. An example of a figure type citation

⚠ **Alert:** You should NOT use *Find Citation(s)* or *Insert Citation(s)* command for inserting figures and tables. Using these commands will only create a text reference in the document without inserting the actual figure or table. You must use *Find Figure(s)* command to be able to insert actual images and tables/charts in the document.

Images from any reference type other than Chart or Table are inserted and formatted as Figures. Images from the Chart or Table reference types are inserted and formatted as Tables.

Some general points regarding working with these reference types in EndNote are:

- Figures are numbered separately from the tables in a document.
- Do NOT use Microsoft Word's Captioning feature in conjunction with EndNote figure citations.
- If your document already contains images that were not inserted using EndNote's *Find Figure(s)* command, those figures will not be included in the figure list. If you are going to use EndNote for figure citations, it is best to insert all figures in a document using EndNote only to ensure accurate numbering of in-text citations and generation of a list of figures.
- Unlike text reference, you cannot disable instant formatting for figures and tables. These reference types are always automatically formatted after you insert them.
- The caption (or title) of the image or table in the document is determined by the value of the text in the **Caption** field in the EndNote reference.

Working with Figures and Tables/Charts

The first step is to make sure that you have stored pictures as a Figure reference type and tables and charts as Chart/Table reference type in an EndNote library. See Chapter 5 for details about creating Figure and Chart/Table reference types.

Inserting Figures or Tables/Charts in a Document

- Place the cursor in your document where you would like the in-text figure or chart citation to appear.
- In Microsoft Word, click *Tools* > *EndNote* > *Find Figure(s)* (Figure 11).
- In the next *Find Figure(s)* dialog box (Figure 20):
  - Enter a search term to identify the image or table.
  - Click *Search*. EndNote gives you a list of Figure and Chart/Table references that match the search term.
  - Highlight the desired reference. Click *Insert*.

EndNote creates an in-text citation and inserts an image/table with a label in the document immediately after the paragraph that cites it. As you will learn in

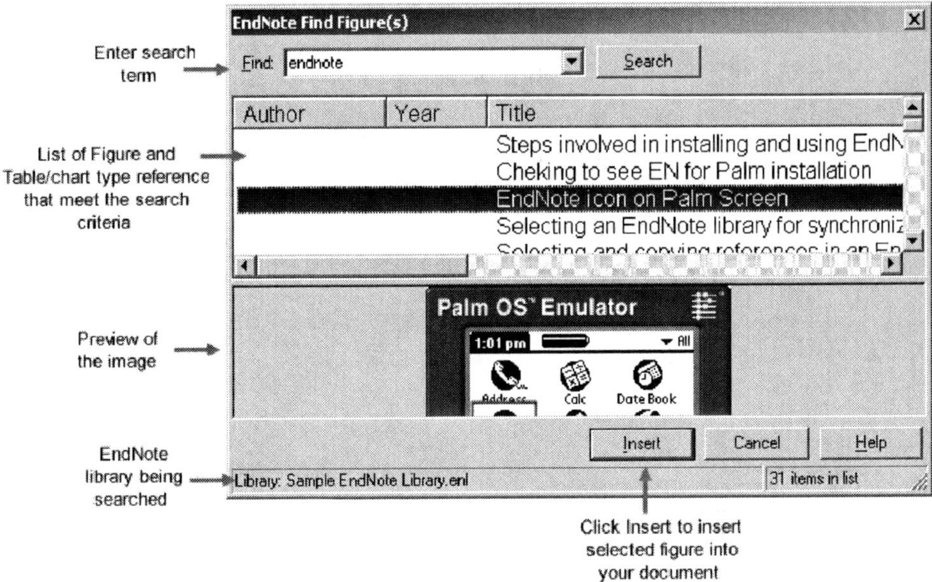

FIGURE 20. Find Figure(s) dialog

the next sections, the placement of figures and tables in a document can be easily modified.

> **Technical Tip:** The *Format Bibliography* command is not applicable to the image and table list. If you add, move, or delete such references, click *Tools > EndNote 7 > Generate Figure List* command to update the in-text citations and figure/table list.

Customizing Figures and Tables/Charts in a Document

You can customize both the placement and captions of figures and tables in a document. The placement depends upon the output style you have chosen. By default, output styles are set up to place figures and tables immediately after the paragraph in which they were cited. You can edit output styles to place figures and tables at the end of the document.

> **Technical Tip:** There is no global editing command in EndNote that will modify all output styles to change figures' and tables' placement setting. You have to change each individual output style you plan to use.

To customize these settings:

- Open an output style for editing (Figure 3). Figure 21 shows Vancouver style editing dialog.
- Click on the Figures heading. Select options for Placement and Captions. Select "Insert figures as a list at the end of the document" if you would like all the images to appear at the end of the document, and not at the place of citation.
- Click on the Tables heading. There are similar options to choose from in this heading as well.
- Click on the Separation and Punctuation heading (Figure 22). This heading allows you to set further preferences, including option to select a new page for each figure and table in the list.
- Close the window and save these settings.

Technical Tip: Choosing different options under the Figures and the Tables heading in the style editing dialog allows you to place figures differently than tables.

Secondly, even if your output style is set to place figures and tables in the text at the end of the paragraph, you may want to move these images to precisely fit your need. However, you would not want this placement to be disturbed during the next execution of the *Generate Figure List* command. To accomplish this, you need to set the CWYW preferences for figures and tables, which will override the settings from the current output style. To set CWYW preferences:

- Click *Tools > EndNote 7 > Cite While You Write Preferences*.
- In the next dialog box, select the Figures and Tables tab (Figure 23).
- Select the second radio button to keep the custom placement of figures and tables.
- Click *OK*.

Sending Paper to Publisher/Sharing with Others

Here are some important concepts pertinent to sharing your manuscripts containing EndNote references.

Field Codes

When you insert a reference in a document using EndNote, EndNote embeds a set of complicated codes in the in-text citation. This code contains complete information about the reference and provides the facility of features such as traveling library

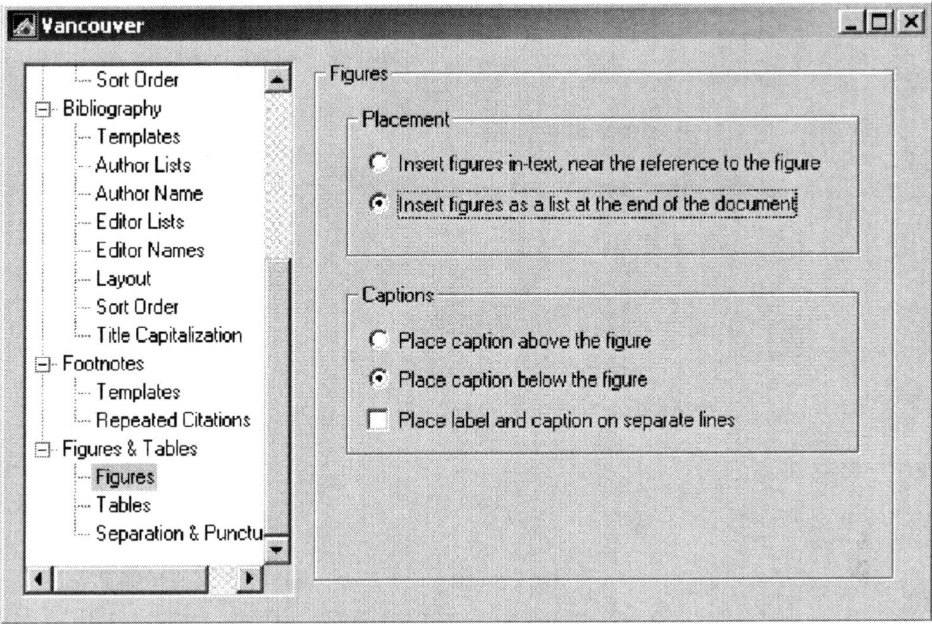

FIGURE 21. Customizing figures and tables in a document

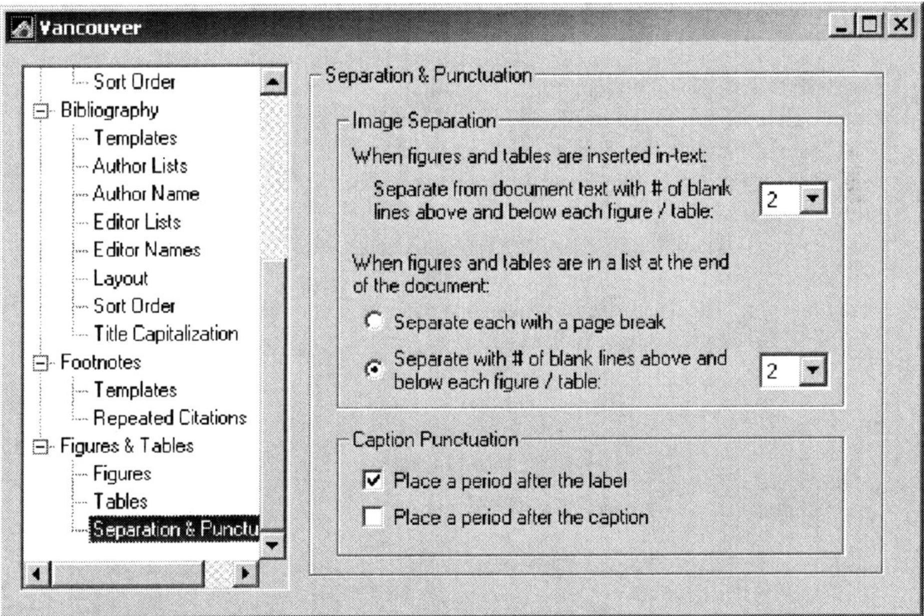

FIGURE 22. Customizing punctuation for figures and tables

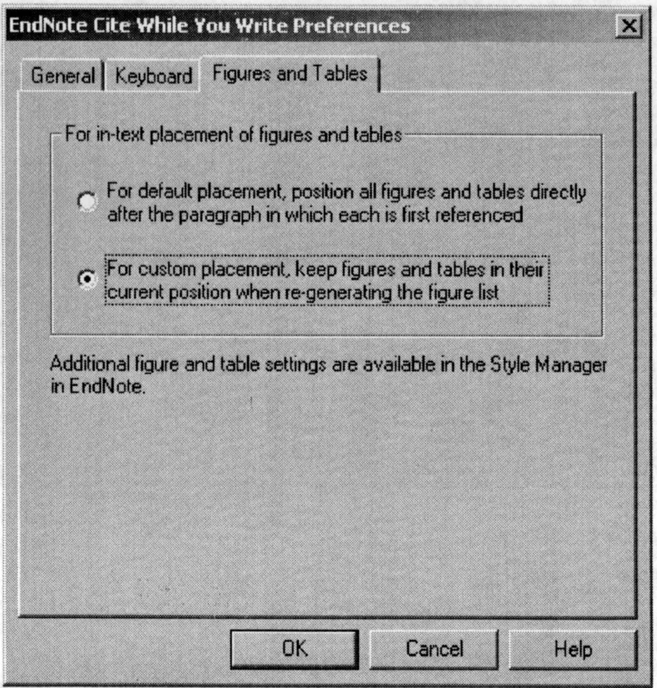

FIGURE 23. CWYW setting for Figures and Tables

(see next section) and reformatting of citations as many times as you wish, even if the original library is not available.

Figure 24 shows how a routine numbered in-text citation and its associated reference appear in a document. If you want to see the field code associated with this citation, place your cursor in the citation and *right-click*. Select *Toggle Field Codes* from the submenu (Figure 25). You should now see the long set of field codes associated with this citation (Figure 26). Seeing this field code is of no direct practical value, but it does give you an idea of the power of field codes. Click *Toggle Field Codes* again to hide the field code. You should NOT modify field codes in a document as doing so may corrupt your references.

The American Psychological Association has published a manual to specify styling and formatting of manuscripts [1].

[1] American Psychological Association. Publication manual of the American Psychological Association. Washington, DC: American Psychological Association, 2001.

FIGURE 24. A citation with field codes hidden

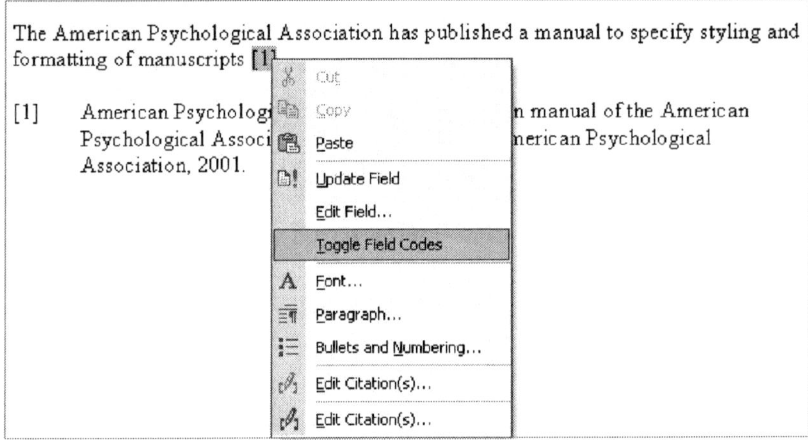

FIGURE 25. Menu to select Toggle Field Codes command

Traveling Library

Embedded within the field codes is, essentially, a 'copy' of the library containing only the references used in the manuscript. This embedded library is called the 'traveling library.'

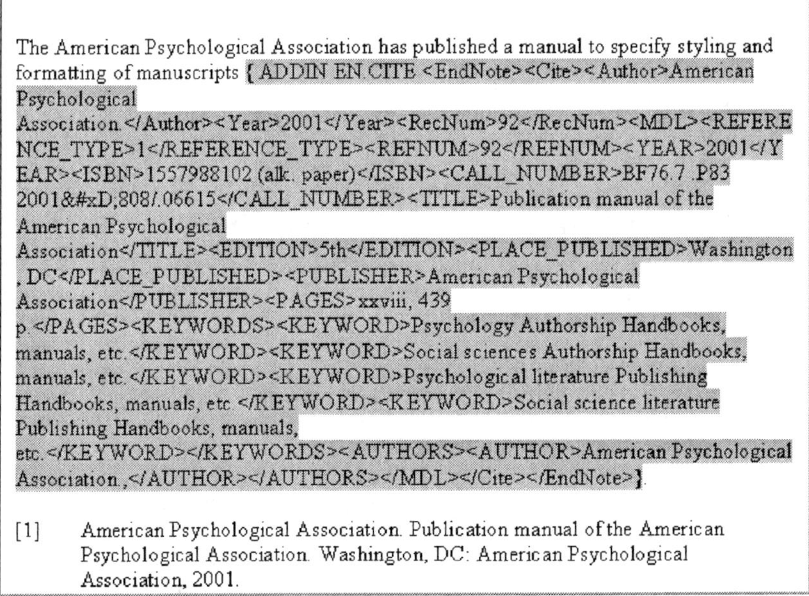

The American Psychological Association has published a manual to specify styling and formatting of manuscripts { ADDIN EN.CITE <EndNote><Cite><Author>American Psychological Association.</Author><Year>2001</Year><RecNum>92</RecNum><MDL><REFERENCE_TYPE>1</REFERENCE_TYPE><REFNUM>92</REFNUM><YEAR>2001</YEAR><ISBN>1557988102 (alk. paper)</ISBN><CALL_NUMBER>BF76.7 .P83 2001
808/.06615</CALL_NUMBER><TITLE>Publication manual of the American Psychological Association</TITLE><EDITION>5th</EDITION><PLACE_PUBLISHED>Washington, DC</PLACE_PUBLISHED><PUBLISHER>American Psychological Association</PUBLISHER><PAGES>xxviii, 439 p.</PAGES><KEYWORDS><KEYWORD>Psychology Authorship Handbooks, manuals, etc.</KEYWORD><KEYWORD>Social sciences Authorship Handbooks, manuals, etc.</KEYWORD><KEYWORD>Psychological literature Publishing Handbooks, manuals, etc.</KEYWORD><KEYWORD>Social science literature Publishing Handbooks, manuals, etc.</KEYWORD></KEYWORDS><AUTHORS><AUTHOR>American Psychological Association.</AUTHOR></AUTHORS></MDL></Cite></EndNote>}.
[1] American Psychological Association. Publication manual of the American Psychological Association. Washington, DC: American Psychological Association, 2001.

FIGURE 26. A citation with field codes showing

> ⚠ **Alert:** The reference data saved in field codes for each citation includes all reference fields except Notes, Abstract, Image, and Caption. Therefore, the traveling library does not contain Notes, Abstract, Images, or Captions.

Sharing Your Document with Others: Creating a Traveling Library

If you send a Word document with EndNote linked references to a colleague without sending the associated EndNote library, your colleague can still have access to all the references because of the traveling library feature. Not only that, your colleague can create his or her own copy of the EndNote library by importing these references from the document. This is a very useful feature, especially if you are collaborating with multiple authors on a manuscript. It eliminates the need for sending EndNote libraries to these various authors.

There are two ways to create an EndNote library from a document. There is no difference between the two and you can use either one of them.

(A) From Word

- Open the document in Microsoft Word.
- Click *Tools* > *EndNote 7* > *Export Traveling Library* in Microsoft Word (Figure 11).
- In the next Export Traveling Library dialog box, select the option of either adding these references to an existing library, or of creating a new library.

(B) From EndNote

- Open both the Word document with references and EndNote.
- Click *Tools* > *CWYW* > *Import Traveling Library* in EndNote (Figure 27).
- The same dialog as above opens with the same options.

> **Technical Tip:** There is another use for traveling library that I find quite helpful. If I have a large EndNote library and I quickly want to get a list only of the references I used in a paper, I simply create a traveling library from the paper and save it as a new library.

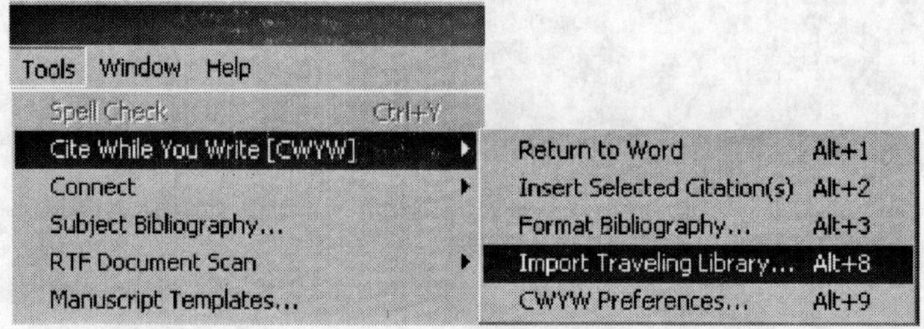

FIGURE 27. Creating Traveling Library from EndNote

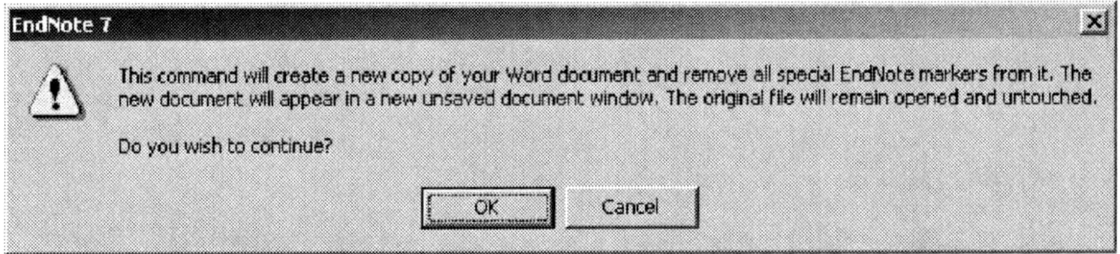

FIGURE 28. Remove Field Codes dialog box

Sending Your Paper to a Publisher

Publishers sometimes expect authors to submit an electronic copy of a paper in addition to, or instead of, a paper copy. Before you submit the electronic copy to a publisher, you must remove the field codes from the paper because the field codes in your document may create incompatibilities with the publishers' software.

Removing field codes saves the formatted citations and bibliography in your paper as plain text. Note that since the field codes do not show up in the printed copy, it doesn't matter whether you remove them or not if you are submitting only a printed copy.

Removing Field Codes

- Open your Word document with EndNote references.
- Click *Tools > EndNote 7 > Remove Field Codes* (Figure 11).
- You will receive an alert dialog box. Click *OK* (Figure 28).
- A copy of the document, without field codes, appears in a new document window. The original document remains untouched.

> ⚠ **Alert:** Once field codes are removed, the paper becomes "unlinked" from the library and you can no longer reformat this new document. In the case of Figure type citations, images in the figure list become GIF files as if they were copied and pasted into the document.
>
> If you are working with master and subdocument in Word, the Remove Field Code command will remove codes from the original documents. You should manually save a copy of these documents and then remove codes from the copies.

Other Tasks

Citing References in Footnotes

To cite references in footnotes, first you have to create a footnote in Word and then use EndNote to insert references in that footnote. Like the in-text citations and bibliography, the format of the reference in the footnote will depend upon the output style you have chosen.

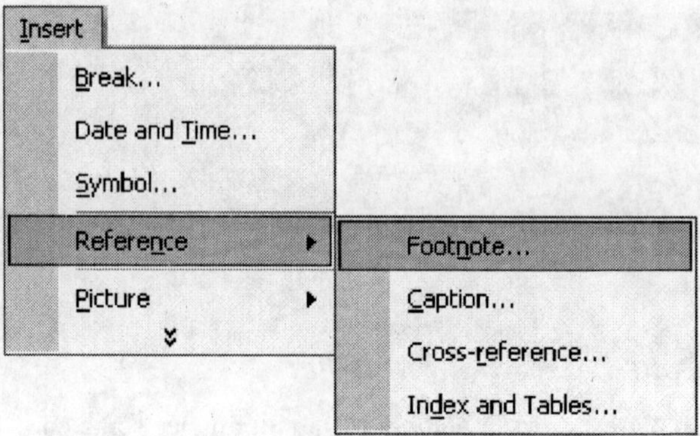

FIGURE 29. Inserting a footnote in Word

(A) First, create a footnote

- In Word, click *Insert > References > Footnote* (Figure 29).
- Select footnote options, such as the location and the format of the footnote, in the next dialog box (Figure 30).

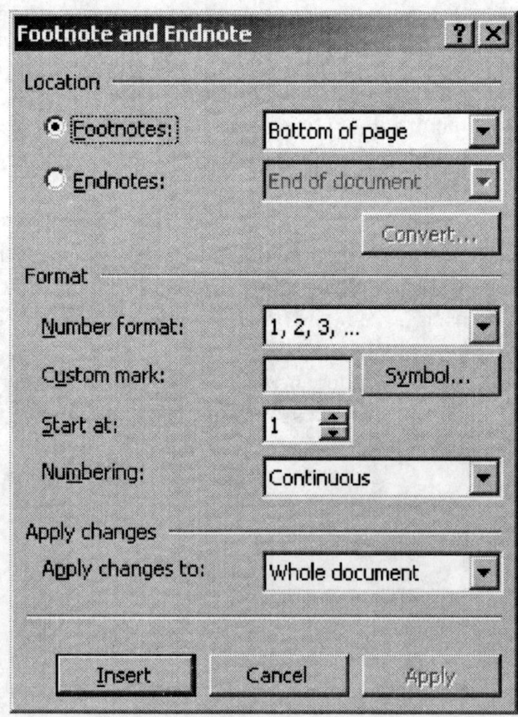

FIGURE 30. Selecting footnote options

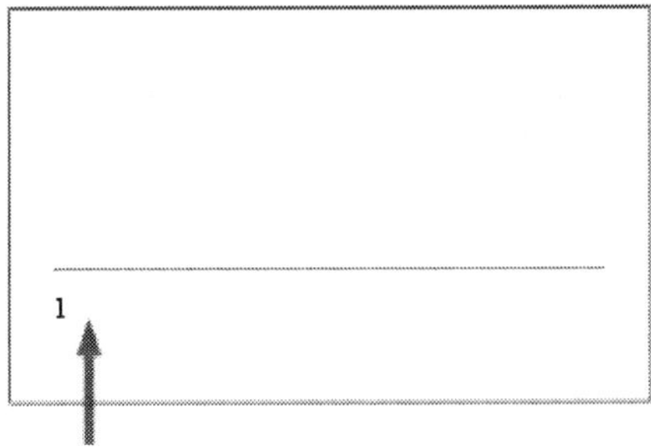

FIGURE 31. Placing cursor in footnote

(B) Then, insert citation(s) in the footnote

- Place your cursor in the footnote (Figure 31).
- Click *Tools* > *EndNote* > *Insert Selected Citation(s)* (Figure 11).
- Click *Tools* > *EndNote* > *Format Bibliography* to format citation(s).

Technical Tip: One problem with citing in footnotes is that by default, EndNote will also include these references in a reference list/bibliography. If you do not wish the footnote references in a separate reference list, you will need to specify this in the output style you are using for your bibliography formatting. To do this:

- Click *Edit* > *Output Styles* > *Open Styles Manager.*
- Select the style you would like to modify. Click *Edit.*
- Select templates under Footnote heading (Figure 32).
- **Uncheck** the box next to "Include References in Bibliography."

Now you should not see the footnote citations in the reference list. Note that you can use this editing screen also to specify the format of footnotes and of repeat citations.

Customizing Footnote Citations

You may want the styling of the footnote in your paper to be different than the styling of the rest of the bibliography. By default, EndNote output styles are set to format citations in footnotes in the same way as the bibliography. To change this, you must modify the footnote section of your output style:

- Click *Edit* > *Output Styles* > *Open Styles Manager.*
- Select the style you would like to modify. Click *Edit.*
- Select templates under Footnote heading (Figure 32).
- Click on the dropdown list to select a style different from the bibliography style for your footnote citations.

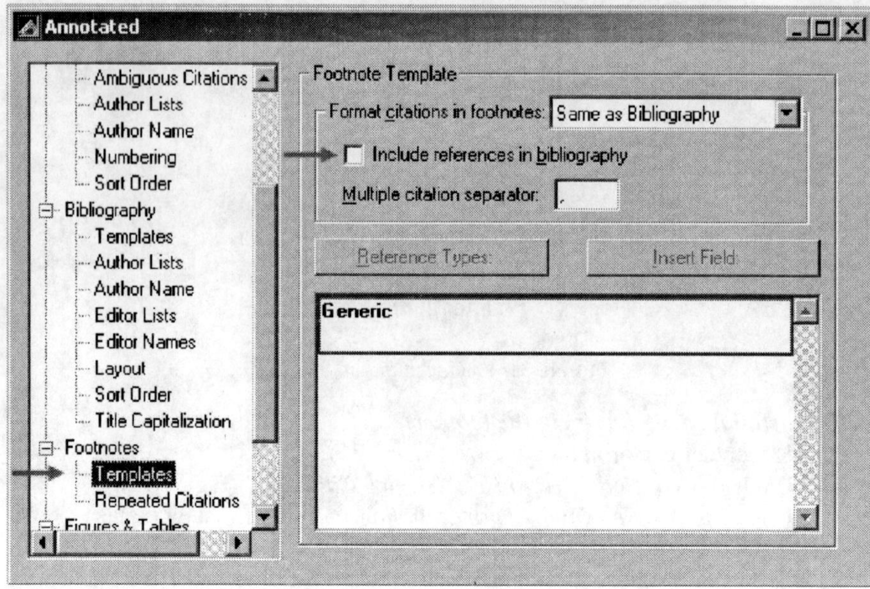

FIGURE 32. Customizing an output style for footnotes

Creating an Independent Bibliography

What is an Independent Bibliography?

An independent bibliography is a reference list or bibliography that is not associated with a paper. Sometimes it is useful to create an independent list of references or bibliography without inserting in-text citations in the text. Examples include listing publications in your curriculum vitae, or preparing suggested reading lists for your colleagues or students. What happens to the figure type or table/chart type references in an independent bibliography? You get a list of captions of figures or charts/tables but not actual images or tables.

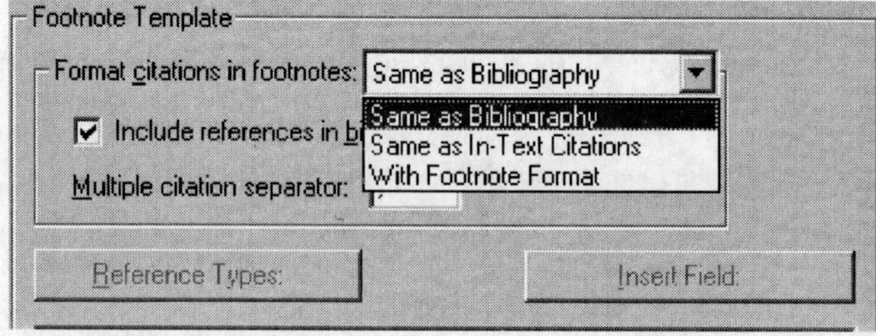

FIGURE 33. Customizing the formatting of footnote citations

Technical Tip: Keep in mind that the appearance of the independent bibliography also depends upon the output style you have chosen. Make sure that you choose an appropriate output style before creating an independent bibliography.

The font of the bibliography is derived from the "General Display Font." See Chapter 4 to learn how to set the General Display Fonts.

Creating an Independent Bibliography

There are four methods for creating independent bibliography. You can choose any one of them according to your preference:

(A) The drag-and-drop method.
(B) The copy-paste method.
(C) The export method.
(D) The print method.

(A) The drag-and-drop method

- Open the EndNote library and a blank Word document.
- Select an appropriate output style (Figure 2).
- Select and drag references from the EndNote library and drop into Word. **Hold down CTRL key as you drag and drop.**

(B) The copy-paste method

- Open the EndNote library and a blank word document.
- Select an appropriate output style (Figure 2).
- Select desired EndNote references.
- Click *Edit > Copy Formatted* in EndNote (Figure 34).
- Click *Edit > Paste* in the Word document.

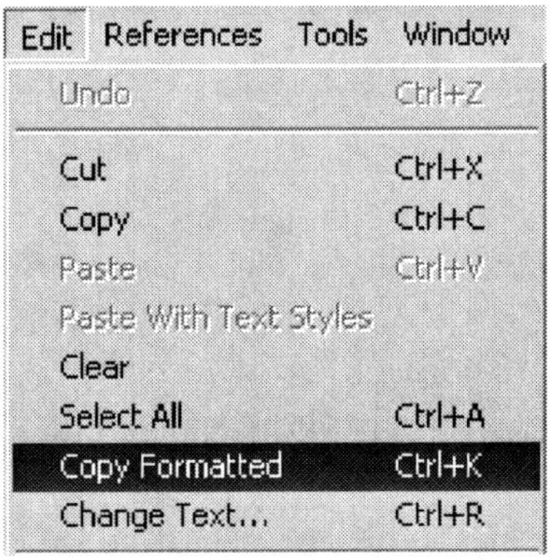

FIGURE 34. The Copy Formatted command in EndNote

(C) The export method

EndNote will export into the following formats: RTF (Rich Text Format), Text, HTML, and XML. EndNote will export all the references that are currently showing in the library window. References are exported in the order in which they are listed in the library window.

- Open the library you would like to export. Make sure only the references you wish to export are showing (see Chapter 6 to learn how to show only selected references).
- Sort references according to your preference by clicking *References > Sort References.*
- Select an appropriate output style.
- Click *File > Export.*
- In the next dialog box
  - Select the file type by picking from the "Save as type" list.
  - Enter a name for the file and click *Save.*

Technical Tip: This exported file can be edited like any other file. Remember if you exported into HTML format, this file is ready to be posted on the web. See Chapter 4 for details about posting EndNote libraries on the web.

(D) The Print method

This method is useful as a quick way of printing selected references directly to paper from your EndNote library without creating an intermediary document like in the previous three methods.

- Open the EndNote library from which you would like to print. Select the references you would like to print (unlike the export method, it doesn't matter what references are showing, the print command affects only the selected references).
- Select an appropriate output style.
- Click *File > Print.*
- Click *Print* in the next dialog box.

Technical Tip:

- References are printed in the order they are sorted in the library.
- If more than one library is open, the print command will apply to the currently active library.
- If you have a reference open, the print command will apply only to that reference.
- Each printed page has 1-inch margins, left-justified text, and a header that displays the library names and page number. You have no option of customizing this appearance. Use one of the previous three methods if you need to customize the appearance of the printed output.

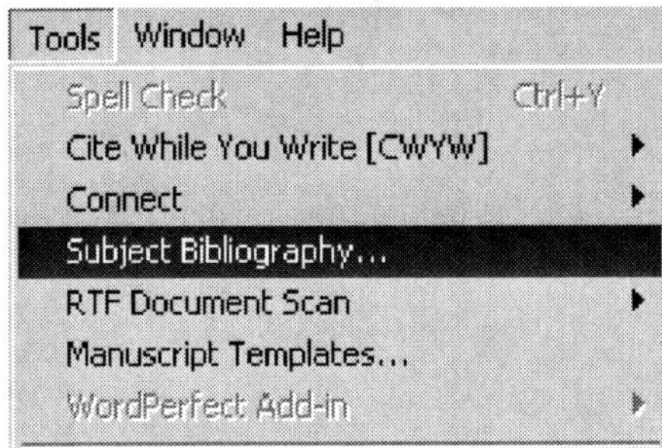

FIGURE 35. The Subject Bibliography command

Creating a Subject Bibliography and Subject List

What is a Subject Bibliography?

A general bibliography is a continuous listing of references. A subject bibliography, on the other hand, categorizes references under distinct headings. These headings could be Keywords, Author names, Year, or any other EndNote field.

EndNote allows you to (a) create subject bibliographies, and (b) create a list of subjects. You can save or print both of these lists.

Note that, like other types of bibliographies, the layout, and format of subject bibliographies also depends on the output style you have chosen.

What is a Subject List?

A subject list is a sorted list of unique terms that occur in particular fields of the records you choose. You can base a subject list on any of the one or more EndNote fields.

Creating a subject bibliography

(A) Generating the subject bibliography

- Open the EndNote library. Select references you would like included in the subject bibliography. To select all references, click *Edit > Select All.* Tip: Even if you select all references, you have the option of selecting only a few keywords of your choice later on.
- Click *Tools > Subject Bibliography* in EndNote (Figure 35).
- In the next window, select the subject fields by which you would like to categorize your references. "Keywords" is the most commonly used category. As you will notice, this listing represents all potentially available reference fields in EndNote. Click *OK* (Figure 36).
- The next screen displays all the terms found in the Keywords field of the references selected.
- Select specific terms as headings, or, if you like, click Select All to create a subject bibliography of all the terms in Keywords field. Click *OK* (Figure 37).

Subject Fields

Selected Fields:

Author
Year
Title
Pages
Secondary Title
Volume
Number
Number of Volumes
Secondary Author
Place Published
Publisher
Subsidiary Author
Edition
Keywords
Type of Work
Date

Select All
Clear Selection(s)
OK
Cancel
Help

List each author separately

In other fields, list each entry that is separated by slash, carriage return, or line feed. (Keywords entries are always listed separately.)

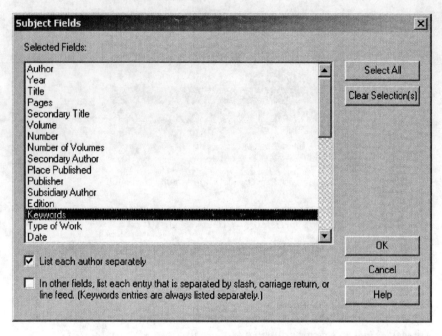

FIGURE 36. Selecting Subject fields

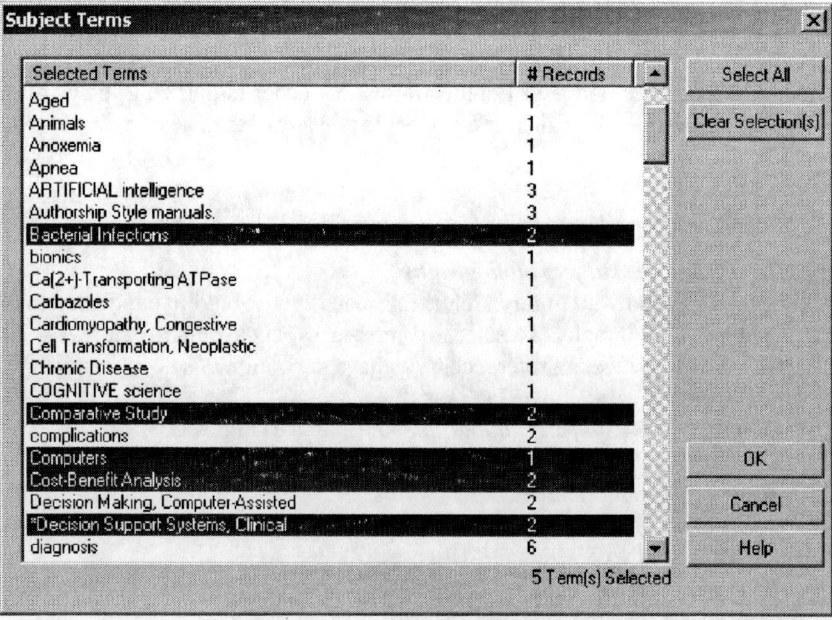

FIGURE 37. Selecting Subject terms

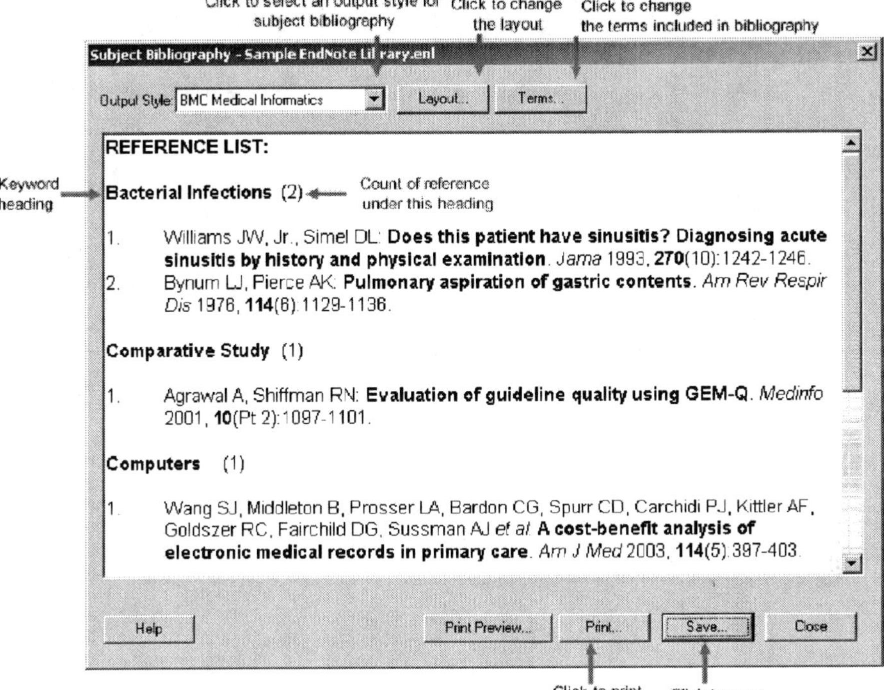

FIGURE 38. A sample subject bibliography

- The next screen shows the subject bibliography generated by EndNote (Figure 38).

The default layout of the bibliography includes a title "REFERENCE LIST" at the top and a list of references formatted in the output style you chose. These references are organized by the terms you chose in the Keywords selection screen. Next to each keyword heading is the count of references under that heading.

(B) Customizing the subject bibliography

If you would like to change the configuration of the default subject bibliography generated above:

- Check the output style you have selected and, if needed, change it to another output style.
- Click the *Layout* button (Figure 38). In the next configuration screen, you will see four tabs – "References," "Terms," "Page Layout," and "Bibliography Layout." Click on these various tabs to bring selection screens to customize the bibliography to your preferences (Figure 39).

(C) Saving or printing the subject bibliography (Figure 38)

Once your subject bibliography is set up as you wish, you can do one of the following:

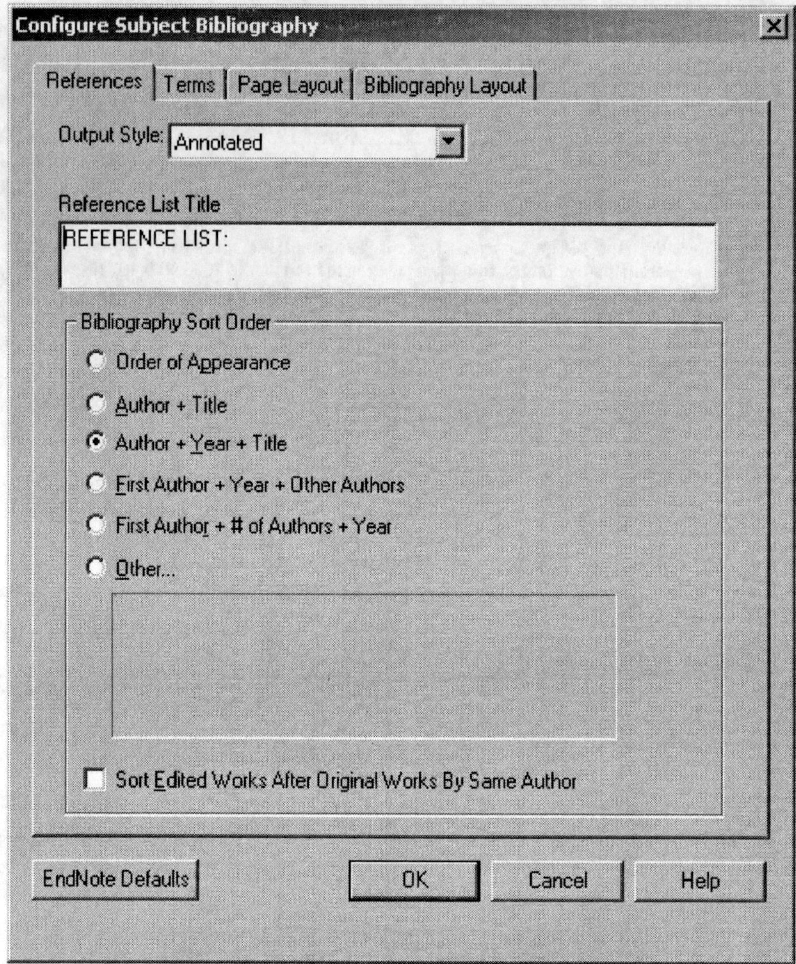

FIGURE 39. Customizing subject bibliography

- Click the *Save* button to save it as a file. EndNote allows you to save this file in the RTF, Text, or HTML formats.
- Click the *Print Preview* button to display a formatted page view.
- Click the *Print* button to print this to a printer.
- Click the *Close* button to close the window.

Creating a Subject List

To create a subject list, you follow essentially the same steps as above, except, during the configuration, you configure the subject bibliography just to choose the terms to display and print.

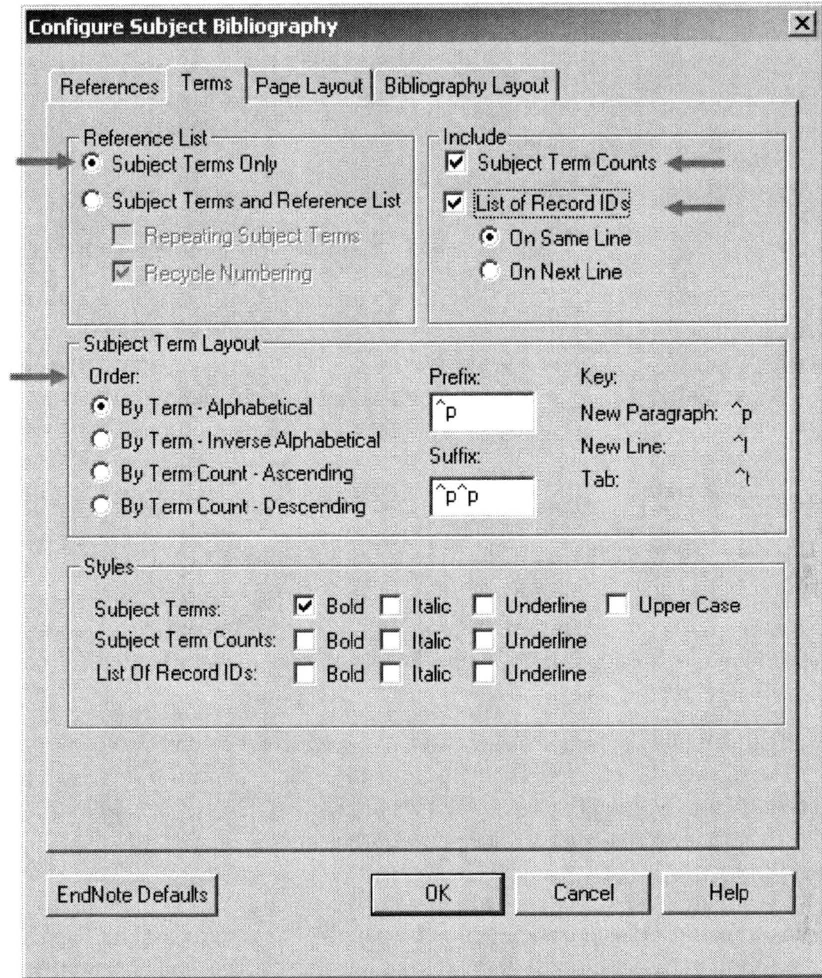

FIGURE 40. Creating a subject list

To do this, make the following changes in the configuration screen:

- After a subject bibliography has been generated, click the *Layout* button (Figure 38).
- In the next screen, click the References tab. Change the title to SUBJECT LIST.
- Click the Terms tab (Figure 40).
  - Set the Reference List option to Subject Terms only.
  - Set the Include option to "Subject term counts" and/or "List of record IDs"
  - Select an appropriate Subject Term Layout
  - Leave Prefix as blank.
  - Type a caret-p (^p) in the Suffix box. This will cause each term to print as a new paragraph on a new line.

FIGURE 41. Accessing CWYW preferences from EndNote

- Click *OK* to save changes. Now you will see a list of subject terms in the subject bibliography window. You can save it or print it in the same way as a subject bibliography.

Setting CWYW Preferences

There are two ways to access and set CWYW preferences:

(A) From EndNote using *Tools > Cite While You Write > CWYW Preferences* (Figure 41).

(B) From Word using *Tools > EndNote 7 > Cite While You Write Preferences* (Figure 11).

Both will bring the same dialog and will implement the same changes.

The CWYW Preferences Dialog box has three tabs: General (Figure 42), Keyboard (Figure 43), and Figures and Tables (Figure 23). The options in these dialog boxes are self-explanatory and allow you to customize CWYW according to your needs.

The Figures and Tables preferences were discussed earlier in this chapter. The other important customization option I would like to highlight is the *Enable Instant Formatting* option available under the General tab. This function turns instant formatting off or on for new Word documents. When you enable instant formatting, in-text citations and references are formatted automatically after you insert a citation in the document. If it is disabled, then inserted citations appear as temporary citations (such as "{Williams, 1993 #125}") and are formatted after clicking *Tools > EndNote 7 > Format Bibliography*.

> **Technical Tip:** In my experience, it is often better to turn off instant formatting because your computer may slow down each time you insert a citation and it is automatically formatted. In addition, it is easier to look up corresponding references in an EndNote library from a temporary citation than a formatted citation, which is often simply a number.
>
> Note that the *Instant Formatting* function does not affect the insertion of Figure and Chart/Table reference types which are always automatically formatted after insertion.

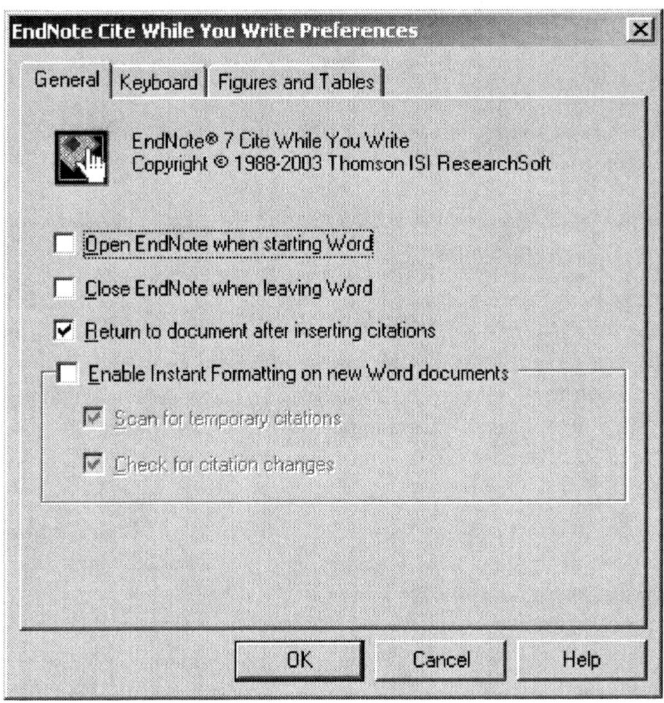

FIGURE 42. CWYW preferences dialog box: General tab

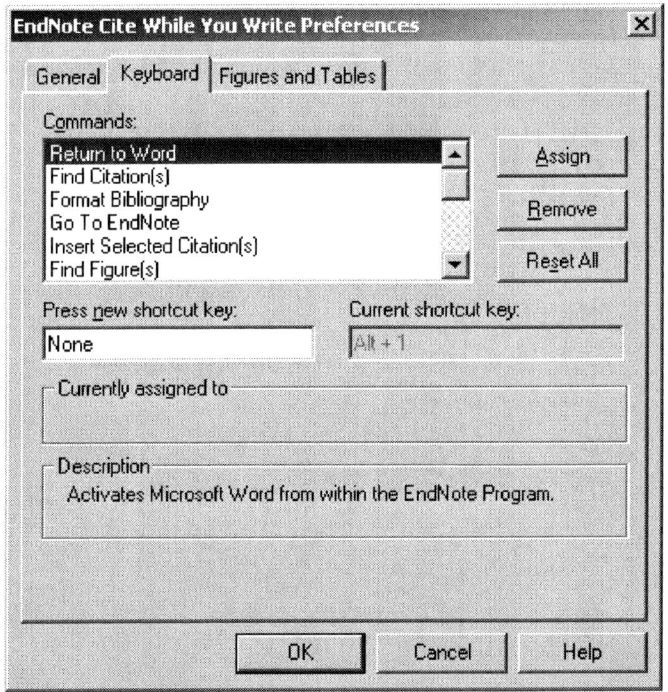

FIGURE 43. CWYW preferences dialog box: Keyboard tab

Miscellaneous Tasks

This section discusses some other tasks pertinent to creating bibliographies that have not been covered elsewhere.

Inserting the Page Number, on Which the Citation Appears in a Manuscript, in a Reference List.

This is particularly relevant for book writing, as some book publishers use a style that uses no in-text citation and all the references (in the form of notes) are grouped at the end of the manuscript by the page number on the manuscript as shown in the example below. Note that "12" is the page number where the citation text "as discussed in a useful review article appears". Obviously, the page number is not static and can change as you edit your manuscript.

> 12, "as discussed in a useful review article." Williams, J. W., Jr., & Simel, D. L. (1993). Does this patient have sinusitis? Diagnosing acute sinusitis by history and physical examination. *JAMA, 270*(10), 1242–1246.

To create a reference list with manuscript page number, you need to use the **Bookmark and Cross-Reference** commands of Microsoft Word. EndNote, by itself, is not able to generate manuscript page numbers; it has no mechanism to determine the page number on which your citation text appears. EndNote, however, will be useful to add the actual reference in the list.

A. Create a bookmark

First, create an invisible bookmark in the manuscript where the text "as discussed in a useful review article" appears. To do this:

- Highlight the above text in the manuscript.
- Click *Insert* > *Bookmark* in Word (Figure 44).

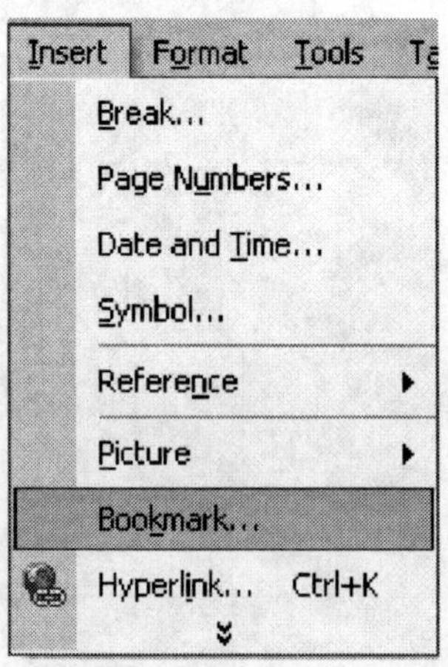

FIGURE 44. Clicking Insert > Bookmark in Word

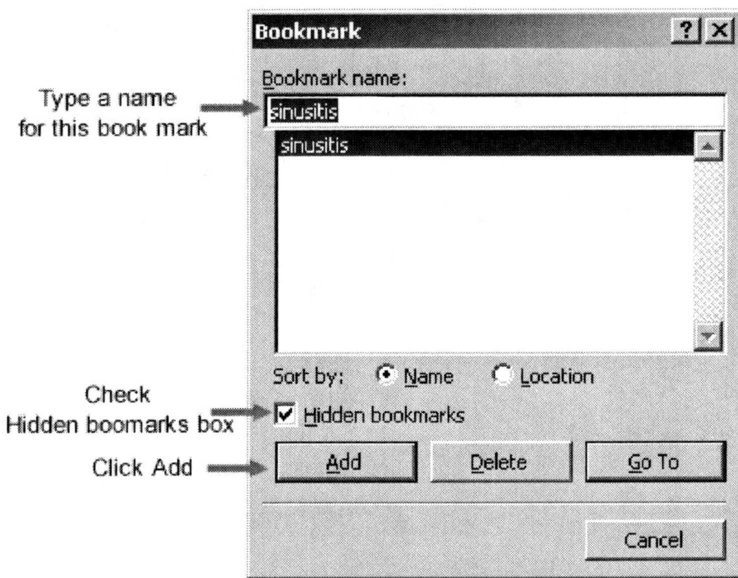

FIGURE 45. Bookmark dialog box

- In the next bookmark dialog box (Figure 45), type a name for this bookmark. This name should be relevant, because you will need it when you create a cross-reference to this bookmark in the reference list. Make sure the "*Hidden bookmark*" box is checked. Click *Add*.

B. Create a cross-reference

Place your cursor in the reference list where you would like this note to appear.

- Click *Insert > Reference > Cross-reference* (Figure 46).

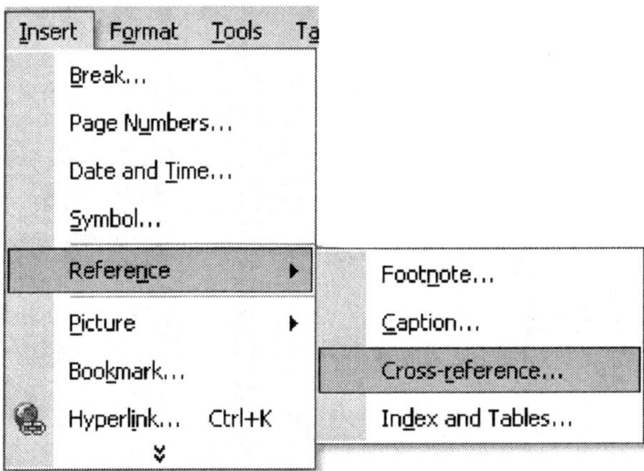

FIGURE 46. Inserting a Cross-reference

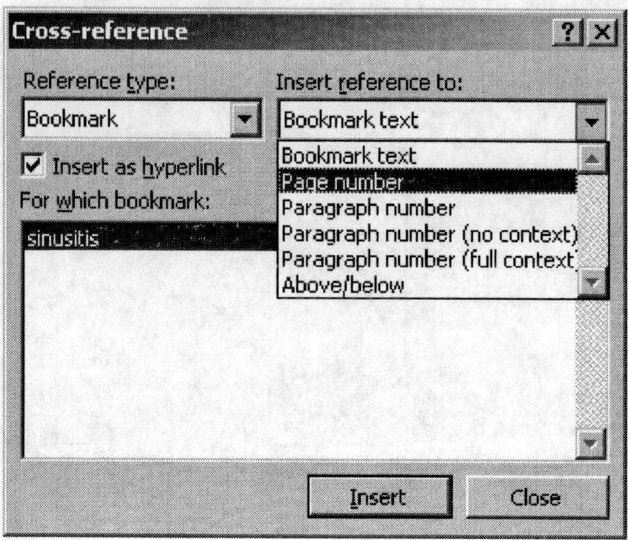

FIGURE 47. Cross-reference dialog box

- In the next cross-reference dialog box, select Bookmark in the reference type dropdown list (Figure 47). Select the name of the bookmark (if more than one). In the "Insert Reference to" dropdown list, first select "Page number," and click *Insert*. Then select "Bookmark text" and click *Insert*.

This should give you a page number and the text of the citation, which will be updated automatically if the text is moved to another page in the manuscript. Now

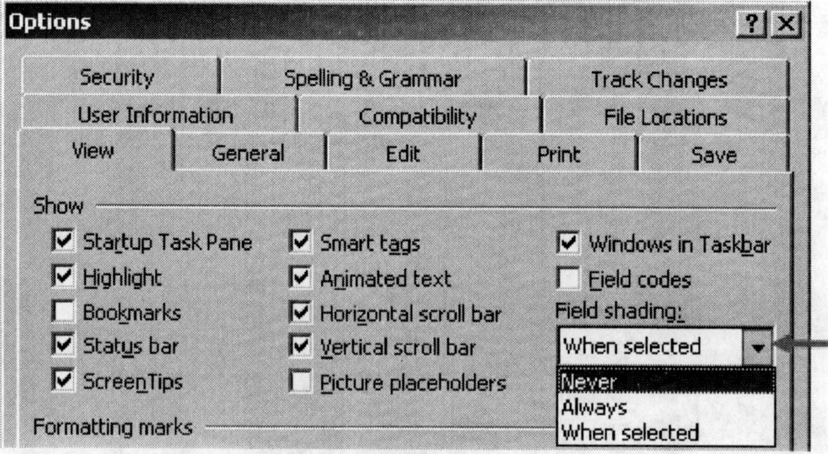

FIGURE 48. Turning off field shading

you can insert a reference without an in-text citation (see the section on creating independent bibliographies) in this reference list.

Turning Off Field Shading

By default, formatted citations and bibliographies in Word are shaded gray when clicked. This is to indicate that these are special EndNote fields. However, sometimes this shading is undesirable particularly if you are trying to edit in-text citations or bibliographies. You can turn off or alter this shading behavior by performing the following steps in **Word (not EndNote):**

- Click *Tools > Options.*
- Select the *View* tab.
- Locate the Field Shading dropdown menu. Select the desired option for field shading. Select "Never" if you don't want field shading to appear at all (Figure 48).

9 Citing References from Sources on the Internet

Things You will Learn in This Chapter

- The differences between citing references from Internet sources (used interchangeably with online or electronic sources) and print sources
- General principles for citing references from Internet sources.
- Examples of how to cite references from various Internet sources such as websites, online magazines and periodicals, e-mails, and web discussion groups.
- How to use EndNote to create a bibliography of references from Internet sources.
- The spelling and definitions of commonly used terms relevant to Internet sources.

Introduction

The Internet is a relatively new, but increasingly important, medium of knowledge dissemination. There is a tremendous amount of information, including biomedical information, available on the Internet, and often you will find that you need to give reference to a document, white paper, or a report that is available from an Internet source. Some examples of reference material available on the Internet include:

- Articles also available in the print media such as scientific journals, newspapers, or newsletters.
- Online material not available in the print media such as research papers, government reports, and online journals or books.
- Documents with a quintessentially Internet format such as websites, e-mails, or newsgroup messages.

The Internet is a very convenient source of scientific material for professionals. Online sources can be retrieved instantly at any time and from any location.

Information on the Internet is often more up-to-date than print information. In addition, searching on the Internet is faster, easier, and more direct than looking for a print journal or book in a library with thousands of paper volumes. This chapter discusses various aspects of citing reference material you retrieve from the Internet.

What's Different about Citing Internet Sources?

As discussed in Chapters 1 and 8, a variety of style manuals provide guidelines for formatting the elements of bibliography and in-text citations from print sources. However, there are still no clear standards for styling references from Internet sources. Major challenges in citing Internet sources are:

Critical Citation Elements may be Missing

- Page numbers: Pagination, an important element of print publication, has little or no meaning in online material. Web pages and e-mails are not broken up into standard, numbered page units, as are books or journal articles. A web page is usually one page, regardless of the length of the material.
- Author/Editor: It may be difficult to identify the author of the information on a website. This has to do, fundamentally, with the differences the between the publishing processes of conventional print medium and the Internet. In the print medium, written material typically undergoes a process of peer-review and/or editorial review; however, websites are often published with little editorial oversight and no identifiable information about the author.

Date of Publication is not Available or not Static if Available

When an article is published in a printed book or a journal, it carries clear, unambiguous, and inalterable information about the date of the publication of the material. Websites often do not provide a clear date of publication. A document on the web can be altered or deleted from the website at any time without leaving any trail of editing, which makes information about the date of publication very difficult to obtain.

Therefore, a new element is generally added to citations from Internet sources: the date when you accessed the source. Since web pages are so volatile, the access date provides a point of reference and may be the only means of designating the specific "edition" of an online document. Reporting where a document was located on a certain date does not guarantee that the document will not be revised or moved, but it does lend credibility to the work of the writer citing the document.

The Location of Information is not Static

If you want to retrieve an article from a print journal, you need to know the volume, issue, and the date of publication of the journal. Similarly, the location of a web page is identified by its URL (short for Uniform Resource Locator), which is akin to its address on the Internet. In contrast to the static nature of the location of a printed material, the URL for a web resource can be changed at anytime.

General Principles for Citing Internet Sources

A huge variety of material exists on the Internet with no uniformly agreed format for structure and presentation of the content. Regardless of format, authors citing Internet sources should observe the following general principles:

- Even if complete information about the reference is not available, provide as much information as you can, especially about the URL and the date you accessed the reference online.
- If a publisher does not provide specific guidelines to cite electronic resources, draw an analogy to how you would cite a print article and include similar elements in the electronic citation as well.
- Place special emphasis on the accuracy of the URL as it is critical to finding the reference online. The next section discusses some strategies for accurately citing URL in your manuscript.
- The National Library of Medicine recommends including the word "Internet" in brackets after the title to make it clear that the item being cited is derived from an online source. For example:

> "Patrias K. National Library of Medicine recommended formats for bibliographic citation. Supplement. [Internet]. Bethesda, MD: National Library of Medicine, 2001."

Generally at a minimum, an Internet source reference should provide the following elements (the arrangements of which may differ according to the bibliography output style you are using):

- Document title or description.
- Access date and either the date of publication or update.
- URL.

When in doubt, it may be better to give too much information than too little.

Guidelines for Citing Internet Sources

Here are some general guidelines for citing various elements of a reference from Internet sources:

Authors

As you may have observed, many websites do not carry easily discernable information about the author:

- Look at the top and the bottom of the web page as the likely places where author information may be found.
- Look for "About this site" or "About us" page on the website, if available. This page may have more information about an author or editor of the website.
- Many websites have the name or e-mail address for the "web master" of the site. This person is generally the person who created the technical design of the site and is probably not the author of the information on the website.
- If there is no identifiable author, do not use "anonymous"; instead, leave the author field blank.

Title

If there is no easily identifiable title for the reference on the Internet, here are some basic rules to identify the title (Figure 1):

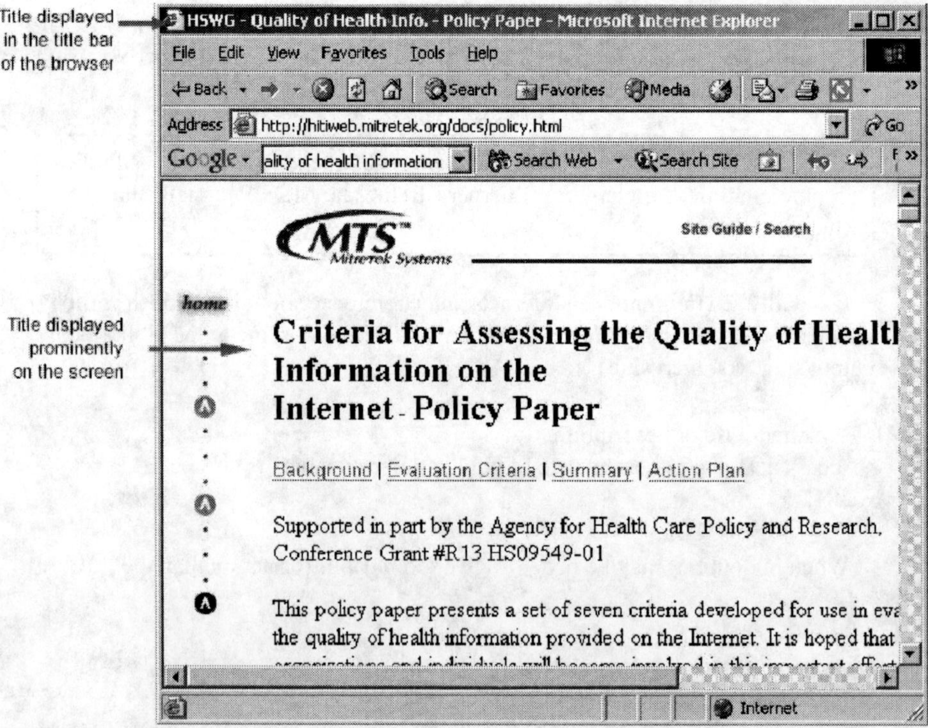

FIGURE 1. Finding the title of a document on the Internet

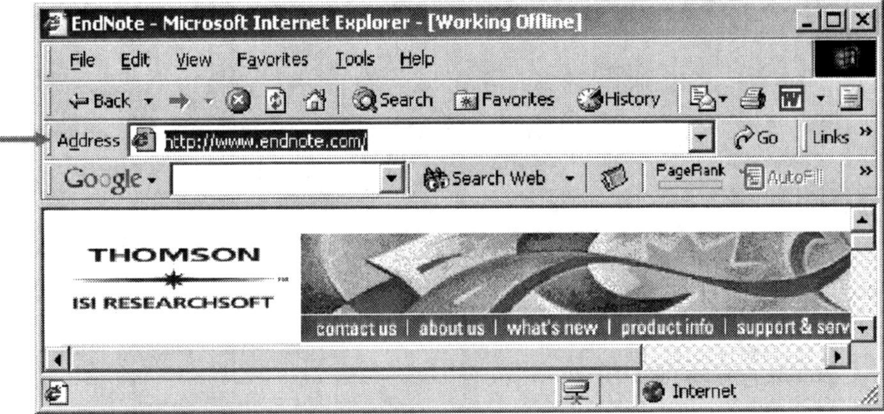

FIGURE 2. Highlighting URL in the browser's location space

- Look for a title for the web page in the title bar of the web browser (generally in the top left corner).
- Look for what is the most prominent (usually the largest) wording on the screen.
- Look for wording followed by a copyright or registered trademark symbol (© or ™).

If a title cannot be determined, you may consider constructing a title by using the first series of words on the screen as a title or leave the title field blank.

URL

The URL is the most critical element of the citation of an online source. If the URL is not accurate, readers won't be able to find the reference, which will diminish the credibility of your manuscript. The easiest way to write the URL of a web page accurately is by clicking in the browser's location space to highlight the URL (Figure 2). Then click *Edit* > *Copy* (Figure 3). Then paste the URL directly into your paper, or the URL reference field if using EndNote.

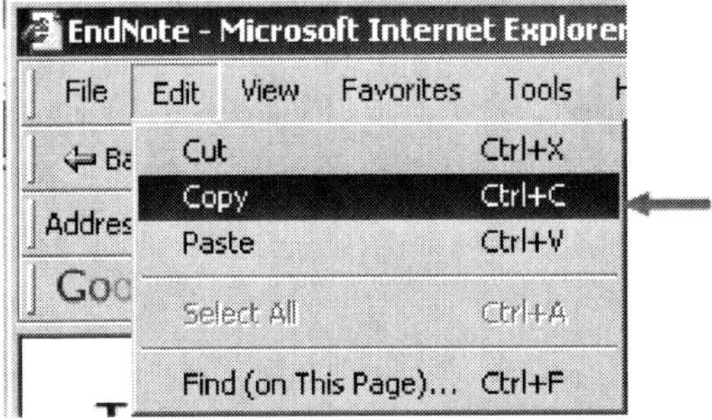

FIGURE 3. Clicking Edit > Copy to copy the URL

> ⚠ **Alert:** A nonfunctional URL is one of the commonest complaints from readers trying to access an online source cited by the author. This may be because the URL is typed incorrectly by the author or the reader. URLs may also fail because the website server is down temporarily or because the publisher of the website has moved or deleted the referred web page from a website.

Some general guidelines for including URLs in Internet citations include:

- Print a copy of the web page you are citing that shows the URL. Since Web pages tend to disappear, it's a good idea to save or print a copy of documents that you cite, so that you can refer to them later. Most web browsers will automatically print the URL of the document and the date of access on the web page print-out.

> **Technical Tip:** After you insert a URL in a Word document, often Word will automatically underline the URL text and change the color of the URL to blue (for example: http://www.google.com). To revert this URL text to match the rest of the document:
>
> - Right click on the URL
> - Select *Remove Hyperlink*.

- If you are using EndNote, put URL under the URL field of the reference. More information about using EndNote for Internet references is provided later in this chapter.
- The URL should *never* end with a period, unless the URL itself ends with a slash. For example:

CORRECT	**INCORRECT**
• http://nnlm.gov/. • http://nnlm.gov/pnr/news/200107 /locator.html	• http://www.nlm.nih.gov/ medlineplus.osteoarthiritis.html.

- Some websites generate pages from a database. These URLs may be lengthy and complex, such as http://www.ncbi.nlm.nih.gov/entrez/query.fcgi?cmd=Retrieve&db=pubmed&dopt=Abstract&list_uids=15360785. Make sure you transcribe these lengthy URLs accurately into your paper.
- Be sure the URLs in your paper points to the correct web page. Test the URLs in your references regularly while writing your manuscript. Test them again when you submit it for review and finally for publication.
- Represent the URL accurately, with no added punctuation or spacing. If it is necessary to divide the URL between two lines, break only after a slash mark and do not insert a hyphen at the break.

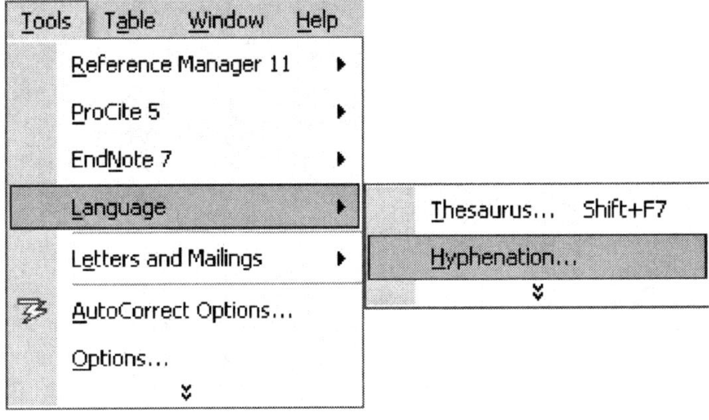

FIGURE 4. Selecting the Hyphenation command

Technical Tip: Microsoft Word may be set to treat the entire URL as a single word and therefore, it will automatically insert hyphenation if a URL is divided between two lines. Make sure to turn off the automatic hyphenation feature of Word if you will be using URLs in a document. To do this:

- Click *Tools > Language > Hyphenation* in Word (Figure 4).
- **Uncheck** the "Automatically hyphenate document" box (Figure 5).

Edition

Most online publications do not have an edition identifier. Some with print counterparts will say "Internet edition." Other words used to express edition in the electronic world include "version," "release," "level," and "update," such as "version 5.1" or "third update." If you find any indication of edition in an online reference, it is useful to include it in reference.

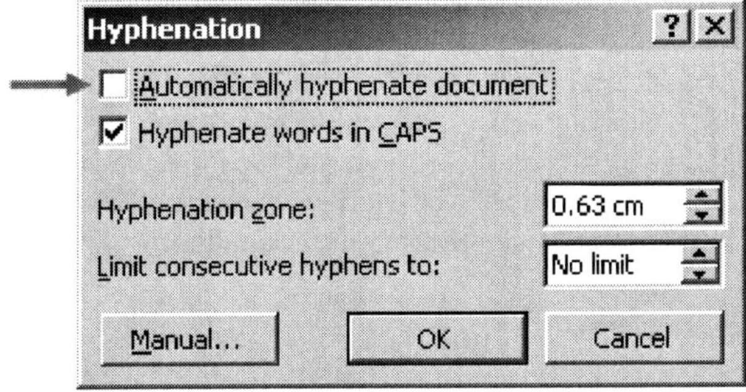

FIGURE 5. Turning off automatic hyphenation in Microsoft Word

Dates

Dates are *extremely* important since the electronic environment is so volatile. The most important date to include in your reference is the access date, that is, the date you viewed, downloaded, or printed references from the online source. You should also try to include the date of publication plus the date of any revisions, if one or both can be found.

Page Information

As discussed in the previous section, the information about pagination is often absent in the material available on the Internet. To give readers an approximate idea of the size of the referred material, you may include an estimate of the online material's size (e.g. 1.5 MB file) in the reference, if available.

In-Text Citations of Internet Sources

For a print source, the exact formatting of in-text citations depends upon the style you have chosen for your paper, for example, the APA 5th style recommends including the author's last name, the date of publication, and/or the page number of the reference in an in-text citation. For Internet sources, some or all of these elements are often missing, making it challenging to create properly formatted in-text citations. Some general guidelines for creating in-text citations for Internet references are:

- Include as much information as is available. If no author's name is available, try to include the file name for example, "cgos.html."
- For references with no publication date, include the access date instead.
- In citations of print sources, you don't always need to repeat the author's name for subsequent citations to the same work. Instead you may simply give a different page number. However, with Internet documents that are not paginated, you may need to repeat the author's name for subsequent citations to the same reference.

Examples of References from Internet Sources

These examples will give you a general idea of the formatting of references from Internet sources. The exact styling of Internet references, just like any other reference, will depend upon the publisher of your manuscript.

Sources on the World Wide Web (WWW)

The following are some sample references from a variety of WWW sources:

Online Journal Article

O'Connor PJ, Desai JR, Solberg LI, Rush WA, Bishop DB. Variation in diabetes care by age: opportunities for customization of care. BMC Fam Pract [Internet] 2003;4 (1):16. Available from <http://www.biomedcentral.com/1471-2296/4/16>. Accessed January 18, 2004

Wick JP, Vernon DD. Visual impairment and driving restrictions. Digital Journal of Ophthalmology, 2002; 8(1). Available from <http://www.djo.harvard.edu/site.php?url=/physicians/oa/280>. Accessed January 18, 2004

Online Magazine Article

Shapiro R. Why is it so difficult to provide universal health care? Slate. <http://www.slate.com/id/2082988>. Published May 15 2003, Accessed January 18, 2004

Online Abstract

Chatham JC. Type 2 diabetes and cardiac dysfunction. 2002. CRISP database. <http://crisp.cit.nih.gov/crisp/CRISP_LIB.getdoc?textkey=6527920&p_grant_num=5R01HL067464-03&p_query=&ticket=6907039&p_audit_session_id=30897186&p_keywords=>. Accessed January 18, 2004.

Online Newspaper Article

Grady D. Jan 13 2004. Heart study prompts call for change. The New York Times. <http://www.nytimes.com/2004/01/13/health/13HEAR.html>. Accessed January 18, 2004.

Online Book

Harnack A, Kleppinger E. 2003. Online! A reference guide to using internet resources. Bedford / St. Martin's Press. Available from <http://www.bedfordstmartins.com/online/index.html>. Accessed January 18, 2004.

Online Book Chapters

Monath, Thomas P. Dengue: the risk to developed and developing countries. In: Roizman, Bernard, editor. Infectious diseases in an age of change: the impact of human ecology and behavior on disease transmission [Internet]. Washington: National Academy Press; 1995, modified 2001Mar 2. P. 43–58. Available from <http://books.nap.edu/books/0309051363/html/43.html#pagetop>. Accessed January 18, 2004.

Web-Sites

The Association of American Medical Colleges website [Internet], Washington, DC. The Association of American Medical Colleges, c. 1995–2003. Available from <http://www.aamc.org>. Accessed January 18, 2004.

Online Databases

OMIM—Online Mendelian inheritance in man [Internet]. National Center for Biotechnology Information. Available from <http://www.ncbi.nlm.nih.gov/entrez/query.fcgi?db=OMIM>. Accessed January 18, 2004.
Cancer query systems [Internet]. Bethesda, MD. National Cancer Institute. Available from <http://seer.cancer.gov/query/>. Accessed January 18, 2004.

Online Graphics

Atlas of Echocardiography [Internet]. New Haven, CT. Yale University; c 1999. Normal ventricular function (Transthoracic view). Available from <http://info.med.yale.edu/intmed/cardio/echo_atlas/entities/index.html>. Accessed January 18, 2004.

E-Mail Messages

Be aware that often e-mail messages are personal communications and it may not be appropriate to include them in reference lists. To document an email message, provide the following information:

- Author's name, recipient's name.
- Date of sending.

- Subject line.
- Date of access.

The following is a sample reference of a fictional e-mail message:

> Doe, John (Center for Medical Studies, Anytown, State. johndoe@email.address). Best ways to cite resources from the Internet. Message to: Doe, Jane (Headquarters, Medical Library, Anytown, State. janedoe@email.address). May 1, 2001, Accessed January 18, 2004. [about 5 paragraphs].

Web Discussion Forum Posting

To document a posting to a Web discussion forum, provide the following information:

- Author's name.
- Date of posting.
- Title of posting.
- URL, in angle brackets.
- Date of access.

The following is a sample reference of a web discussion forum posting:

> Jephat, Chiphamba. Information on the prevalence of smoking in Africa. M. April 30, 2003. Available from <http://www.procor.org/discussion/displaymsg.asp?ref=1242&cate=ProCOR+Dialogue>. Accessed January 18, 2004.

Listserv Message

To document a listserv message, provide the following information:

- Author's name.
- Date of posting.
- Subject line.
- Listserv address, in angle brackets.
- Date of access.

The following is a sample reference of a listserv message:

> Doe, John. The clinical practice guideline for hypertension [Internet]. Rockville, MD. Agency for Health Care Policy and Research, March 13, 1999. Available from <LIST.AHCPR.GOV>. Accessed January 18, 2004 [about 5 paragraphs].

Newsgroup Message

To document information posted in a newsgroup discussion, provide the following information:

- Author's name.
- Date of posting.
- Subject line.
- Name of newsgroup, in angle brackets.
- Date of access.

The following is a sample reference of a newsgroup message:

> Doe, Jane. New trends in graduate medical education. October 1, 2003. Available from <sci.med.education>. Accessed January 18, 2004.

If you cannot determine the author's name, then use the author's email address, enclosed in angle brackets, as the main entry such as:

> <jdoe@e-mailaddress.edu>. New trends in graduate medical education. October 1, 2003. Available from <sci.med.education>. Accessed January 18, 2004.

FTP Sites

To document a file available for downloading via file transfer protocol, provide the following information:

- Name of author or file.
- Date of online publication (if available).
- Title of document.
- FTP address, in angle brackets, with directions for accessing document.
- Date of access.

The following is a sample reference of a file available from an FTP site:

> Gbacc.idx.gz [264 MB], 2004 June 25. <ftp://ftp.ncbi.nih.gov/genbank/>. Accessed 2004 June 30.

Software Programs

Cite the last name and initials of the author(s); the date of publication or release; the title of the software program; the version number; and the publication information.

The following is a sample reference of a software program:

> Polvani KA, Agrawal A, Karras B, Deshpande A, Shiffman R. 2000. GEM Cutter Version 2.0. Yale Center for Medical Informatics, New Haven, CT.

Using EndNote to Manage References from Internet Sources

Now that you are familiar with the general principles of citing references from Internet sources, let us discuss how you can use EndNote to manage such references. Even though EndNote has a reference type called "Electronic Source," it provides sub optimal support for Internet references as you will see in the example below.

The main problem in using EndNote to cite electronic resources is that even if you choose the reference type as "Electronic Source" and input the necessary information, such as URL, Access Date, and Access Year, in various reference fields, the information does not show up in the bibliography. This is because many EndNote defined output styles do not fetch the contents of the URL, access date, and access year reference fields. Remember, the appearance of your bibliography depends upon the output style you choose for your references. Some output styles such as the APA 5th and BMC Medical Informatics are beginning to support "URL" and "Access date" elements of electronic references, but many major styles, including Vancouver, still do not support electronic references in a significant way.

Let us look at an example of how an electronic reference will look in a bibliography using EndNote's default output styles. I have created an "Electronic Source" type reference with the following information (Note: this reference is available in the sample library on the CD):

> **Author:** May, Mike
> **Year:** 2003
> **Title:** Sorting out citation management software
> **Publisher:** The Scientist
> **Access Year:** 2004
> **Access Date:** January 25
> **URL:** http://www.the-scientist.com/yr2003/oct/lcprofile1_031020.html

Using EndNote's default output styles, here is how this reference will look in a bibliography in various styles:

Vancouver Style

1. May M. Sorting out citation management software. In: The Scientist; 2003.

Life Sciences Style

[1] May M. Sorting out citation management software. The Scientist, 2003.

APA 5th Style

May, M. (2003). *Sorting out citation management software*. Retrieved January 25, 2004, from http://www.the-scientist.com/yr2003/oct/lcprofile1_031020.html

CBE Style Manual N-Y

May M. 2003. Sorting out citation management software. The Scientist.

Annotated Style

May, M. (2003). Sorting out citation management software, The Scientist. 2004.

BMC Medical Informatics Style

1. **Sorting out citation management software** [http://www.the-scientist.com/yr2003/ oct/lcprofile1_031020.html]

As you would see from the above example, the APA 5th style outputs most of the elements about the electronic reference. However, this style may not be appropriate for your manuscript. To get the best possible results in citing electronic references using EndNote, decide which output style you need for your manuscript as a whole and then customize it so that your electronic references are formatted correctly as well. Because it is almost inevitable these days that you will be referring to Internet resources in your manuscripts, this exercise is worth the extra time and effort.

Customizing EndNote to Cite References from Internet Sources

This is a relatively simple process, as you will see in the next example showing you how to customize the Vancouver style. A similar process can be used to customize any output style for electronic references.

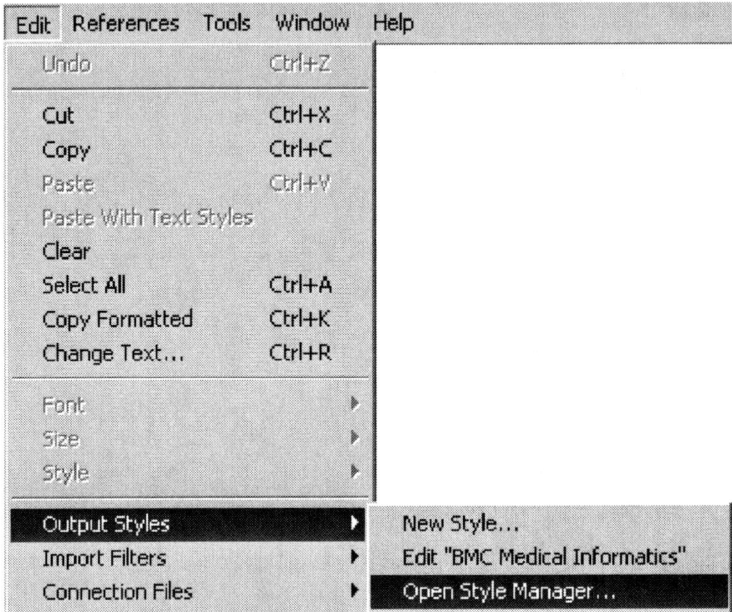

FIGURE 6. Opening the style manager

- Click *Edit > Output Styles > Open Style Manager* (Figure 6).
- In the next screen, select the Vancouver Style by highlighting it and click *Edit* (Figure 7).
- In the Vancouver style screen, click Templates under Bibliography. Click the Reference Type dropdown list and select Electronic Source (Figure 8).
- As you will notice, the Electronic Source box is empty (Figure 9). It means that this style is not set to fetch any special elements such as URL in the bibliography. This explains why you see unsatisfactory results for electronic sources in the Vancouver style.
- You need to edit this box and put in the reference elements you would like to see in the electronic references in a bibliography. To do this, click the Insert Field dropdown list and insert the desired reference elements. The contents of the Electronic Source box should now appear as shown in Figure 10.
- Exit the Vancouver style by closing the window. Click *Yes* in the next screen to save changes.

Now, in a bibliography using the edited Vancouver style, the same citation as discussed in the previous section will appear as the following:

1. May M. Sorting out citation management software. In: The Scientist, 2003. Retrieved January 25 2004, From http://www.the-scientist.com/yr2003/oct/lcprofile1_031020. html.

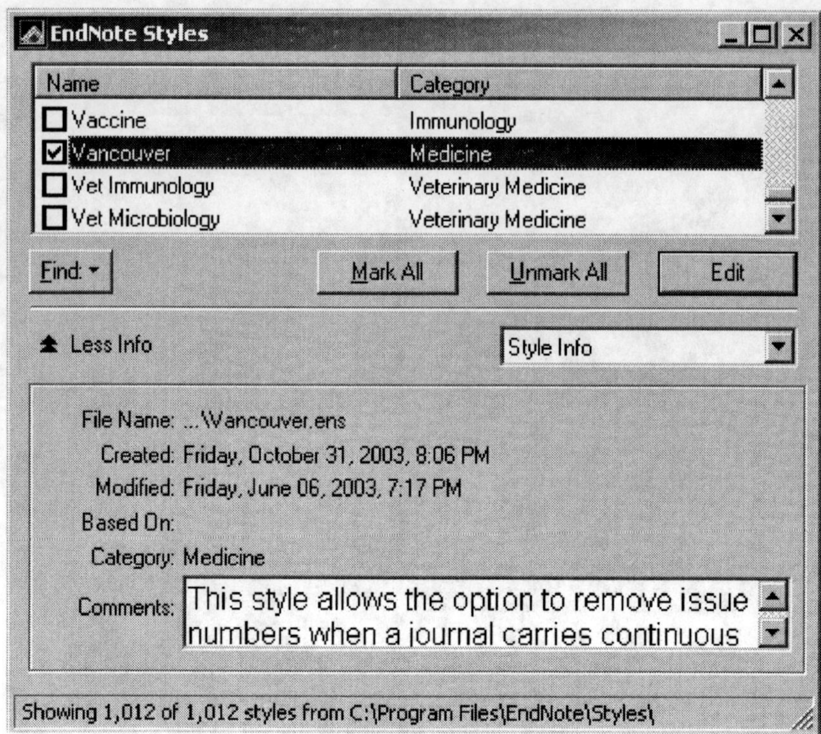

FIGURE 7. Selecting Vancouver style

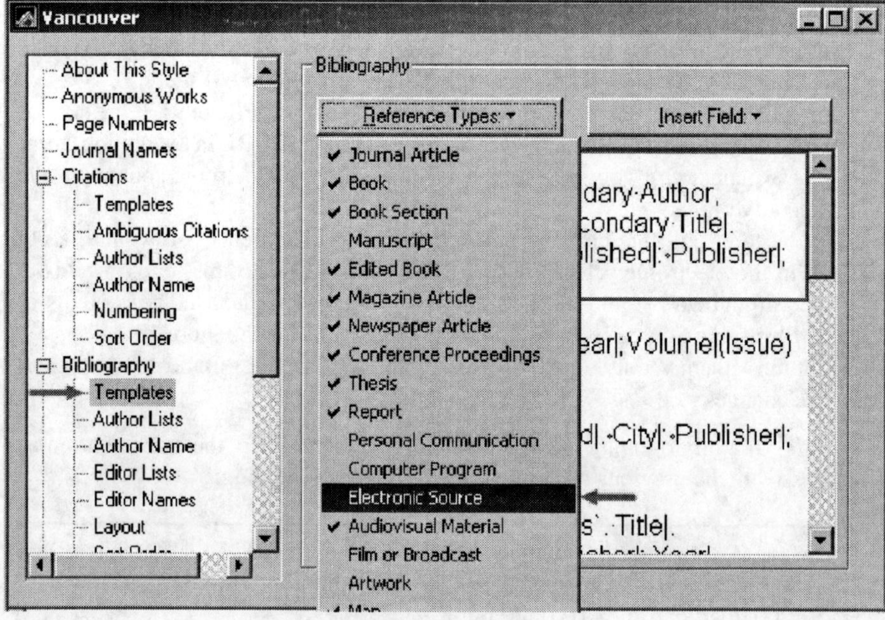

FIGURE 8. Editing Vancouver style

Electronic Source

FIGURE 9. Blank Electronic Source box

Electronic Source
Author. Title. In: Publisher , Year. Retrieved Access Date Access Year , From URL..

FIGURE 10. Edited Electronic Source box

TABLE 1. Spelling and definition of common terms related to online sources

Term and spelling	Definition according to the Merriam-Webster dictonary	Definition according to the Compact Oxford dictionary
URL	the address of a computer or a document on the Internet that consists of a communications protocol followed by a colon and two slashes (as http://), the identifier of a computer (as www.m-w.com) and usually a path through a directory to a file—called also uniform resource locator, universal resource locator	abbreviation uniform (or universal) resource locator, the address of a World Wide Web page
E-mail in Merriam Webster **email** in oxford	1. a means or system for transmitting messages electronically (as between terminals linked by telephone lines or microwave relays) 2. a message sent electronically <sent him an E-mail>	1. noun—the sending of messages by electronic means from one computer user to one or more recipients via a network. 2. verb—mail or send using email.
Internet	an electronic communications network that connects computer networks and organizational computer facilities around the world	an international information network linking computers, accessible to the public via modem links
World Wide Web **Note:** no entry for web site or web page in either dictionary	a part of the Internet designed to allow easier navigation of the network through the use of graphical user interfaces and hypertext links between different addresses—called also *Web*	an extensive information system on the Internet providing facilities for documents to be connected to other documents by hypertext links.
On-line in Merriam Webster **online** by Oxofrd	connected to, served by, or available through a system and especially a computer or telecommunications system <an *on-line* database>; also: done while connected to a system <*on-line* computer storage>	adjective & adverb 1 controlled by or connected to a computer. 2 in or into operation or existence.

(continued)

TABLE 1. *Continued*

Term and spelling	Definition according to the Merriam-Webster dictonary	Definition according to the Compact Oxford dictionary
Floppy disk, also matches **diskette** in both dictionaries	a small flexible plastic disk coated with magnetic material on which data for a computer can be stored	a flexible removable magnetic disk used for storing data.
CD-ROM	a compact disc containing data that can be read by a computer	*No entry in this dictionary*
CD or **Compact disc**	a small optical disk usually containing recorded music or computer data	a small plastic disc on which music or other digital information is stored in the form of a pattern of metal-coated pits from which it can be read using laser light reflected off the disc.
DVD	a high-capacity optical disk format; also: an optical disk using such a format and containing especially a video recording (as a movie) or computer data	abbreviation of digital versatile disc
Source (May 2004)	www.webster.com	www.askoxford.com

Spelling and Definition of Commonly Used e-Terms

Some of you may be relative newcomers to the world of electronic sources. Table 1 lists some common terms relevant to using Internet resources, their spellings, and definitions as specified by the online editions of two standard English dictionaries—Webster, and Oxford.

10
EndNote for PDA Computers

Things You will Learn in This Chapter

- Basic steps in using EndNote on a PDA.
- Technical requirements for using EndNote on a PDA.
- How to install EndNote for Palm application.
- How to copy the EndNote library from desktop to PDA.
- How to work with the EndNote library on PDA.
- How to beam references between PDAs.
- How to view statistics about the PDA EndNote library.
- How to customize the PDA EndNote library.

Introduction

Personal Digital Assistants, or PDAs, are becoming increasingly popular among healthcare and biomedical professionals. According to a recent survey by Forrester Research[1], 46% of physicians reported owning a PDA, along with 29% of pharmacists, 18% of nurses, 17% of psychologists, and 8% of non-healthcare workers. A recent article in the NY Times[2] predicted that PDAs "may someday be as ubiquitous as the stethoscope."

PDAs increasingly are being used for a variety of applications, such as drug reference databases, e-prescribing, and online textbooks. In this chapter, I introduce you to using a PDA to perform reference management tasks with the help of the EndNote program.

EndNote software (regardless of whether you purchased it on a CD-ROM or downloaded from the Internet) contains two programs:

---

[1]Consumer Technographics 2003 North American Benchmark Study. Available at http://www.forrester.com.

[2]For the doctor's touch, help in the hand. The New York Times, August 22, 2002

1. The main EndNote program for desktop computers. This is the program you have been using so far.
2. The EndNote program for PDAs, known as the 'EndNote for Palm®. This is the program discussed in this chapter.

Just as you enter and save references into the EndNote library on your desktop, you can enter and save references in a PDA. You can synchronize the library between the desktop and the PDA using the standard HotSync™command. Synchronizing EndNote libraries is similar to synchronizing other PDA applications such as address book and calendar.

Technical Tip: PDAs (Personal Digital Assistants) are handheld computers containing at least some basic applications such as date book, address book, task list, and memo pad. A salient feature of all PDAs is the ability to synchronize information with the desktop computer. In addition to providing easy information back up, synchronization allows you to enter data such as an address or an appointment in a desktop computer and copy it onto the PDA.

Basically, PDAs employ two different operating systems:

A. PDAs based on the Palm® operating system. For example, devices by Palm, Inc., Handspring, Qualcomm, and TRG.
B. PDAs based on Microsoft's PocketPC® operating system. For example, devices by Hewlett-Packard and Compaq.

Using EndNote on a PDA provides the benefit of portability by allowing you to carry your references between home and office or while traveling. If you are making a trip to the library to get selected books or journal articles from your reference list, you can take this list with you in a personal digital library on a PDA. You can also electronically share references with your colleagues by 'beaming' them to another PDA.

Figure 1 illustrates an overview of the steps involved in installing and using EndNote on a PDA.

Technical Requirements for Using EndNote on a PDA

- The PDA must use the Palm® operating system (OS) version 3.1 or higher.
- At least 4 MB RAM on the PDA.
- You will need a Serial or USB port on your desktop computer for HotSync operation.
- You will need a HotSync cradle or cable (generally included with the PDA).

At present, EndNote for Palm application is compatible with Palm OS-based devices only, and not with PocketPC-based devices such as those manufactured by Compaq or Hewlett-Packard.

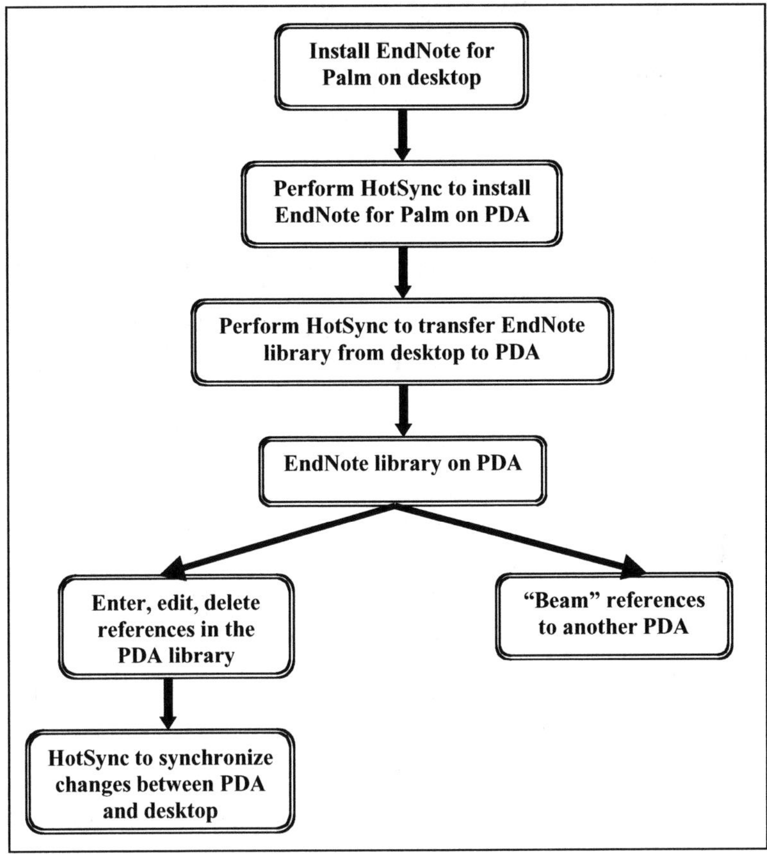

FIGURE 1. Steps involved in installing and using EndNote on a PDA

Another hint of caution: The EndNote vendor also advises that the EndNote for Palm was designed to run on PDAs manufactured by Palm Inc., only. It may be possible to install and run it on PDAs with the Palm OS manufactured by other vendors (e.g., Handspring or Sony), but the EndNote vendor does not recommend or support doing so.

Some example of EndNote compatible PDAs manufactured by Palm, Inc. are:

- All models from the Zire series.
- All models from the Tungsten series.
- m series: m125, m130, m500, m505, m515.
- Palm Vx.

Personally, I have used the EndNote for Palm application on a Palm III xe model without a problem.

Installing EndNote for Palm Application

The installation involves two steps:

Step 1. Install EndNote for Palm on the desktop.
Step 2. Perform HotSync operation to install EndNote for Palm on the PDA.

> **Technical Tip:** Before you begin installing EndNote on your PDA, perform a test HotSync operation to make sure you are able to send files (such as an address book, or datebook entries) between desktop and PDA. If this succeeds, you are ready to install EndNote for Palm application.

Step 1. Install EndNote for Palm on the Desktop

Scenario A. You first installed Palm software on your desktop and are then installing EndNote (recommended)

Palm software is the program that is required to use your PDA (regardless of your use of EndNote) with the desktop. It is best to install the Palm desktop software *before* installing EndNote. Then, as you install the main EndNote software, the EndNote for Palm is automatically installed.

To check whether you have the EndNote for Palm automatically installed on your computer:

- Click *Start* > *Programs* > *EndNote*. You should see Setup EndNote for Palm OS® in the EndNote submenu (Figure 2).

Scenario B: You first installed EndNote and then installed your Palm software

You will need to take the extra step to install the EndNote for Palm on your desktop:

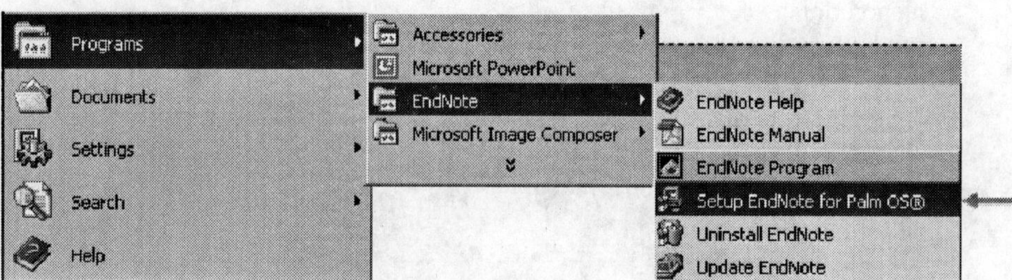

FIGURE 2. Checking the EndNote for Palm installation

FIGURE 3. EndNote icon on the PDA Screen

- Go to folder *Program Files > EndNote > Palm.*
- Double-click on the file ENPalm.exe.
- Follow the instructions to install the EndNote for Palm.

Step 2. Perform HotSync Operation to Install EndNote for Palm on the PDA

After installing EndNote for Palm on your desktop, perform a HotSync operation to copy the application from the desktop to the PDA. You should now see an EndNote icon on the main screen of your PDA (Figure 3).

Copying an EndNote Library from Desktop to PDA

Now that you have installed EndNote on the PDA, you are ready to send a copy of your EndNote library from desktop to the PDA by performing the following steps:

- Designate a *single* EndNote library on the desktop for synchronizing.
  - In EndNote, click *Tools > Configure Handheld Sync* (Figure 4).

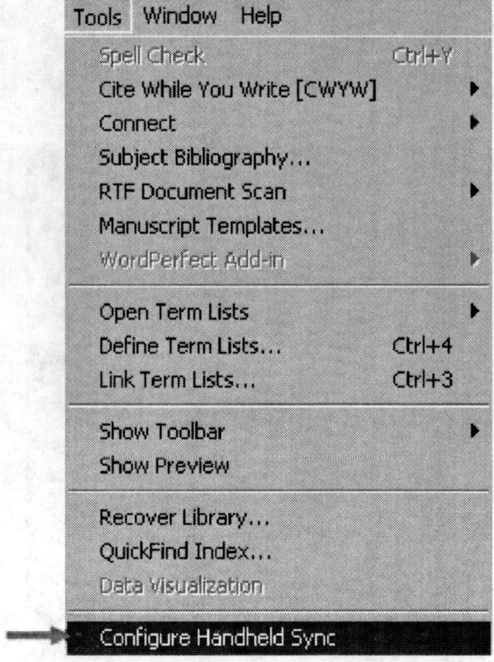

FIGURE 4. Selecting Configure Handheld Sync

- Select an EndNote library from the existing dropdown list or use the *Browse* button to locate EndNote library (Figure 5).
- Click *OK* to save the setting.

• Perform HotSync to copy selected EndNote library to the PDA.

FIGURE 5. Selecting an EndNote library for synchronization

Technical Tip:

- A HotSync operation can **not** be initiated from within EndNote. You must use the PDA cradle or cable to initiate HotSync.
- A HotSync operation will synchronize the EndNote desktop library and PDA library, so that new references will appear in both libraries and edited references will be updated in both libraries. If you edit the same reference in the desktop library and the PDA library, neither record will be overwritten. Instead a second record will be created.

Alert: You can have only a single library in the PDA. If you perform a HotSync operation with a new library, any existing library on your Palm will be replaced.

Copying Only a Limited Number of References to PDA

Normally, you can **not** copy a limited number of references to the PDA; instead the entire designated library is synchronized during HotSync. However, you can do a workaround by copying the selected group of references in a new library and synchronizing the new library with the PDA.

Perform the following steps if you want to copy only a limited number of references, and not the entire library, to PDA:

- Copy selected references to an interim new desktop EndNote library.
  - Click on the references to select. Hold *Ctrl* key while clicking to select multiple references (Figure 6).
  - Select *Edit > Copy* from the menu bar (Figure 7).
  - Select *File > New* from the menu bar to create a new library.
  - Select a name and location for new library.
  - In the new library window, select *Edit > Paste*.

  You should now see selected references in the new library window.

- Designate this new library for synchronization with *Configure Handheld Sync* (Figure 4).
- Synchronize PDA with the new library.

What Happens to 'Figure' Type References in the PDA EndNote Library?

In the PDA EndNote library, you will see only the name of the image in the image field, and not the thumbnail icon for the image (Figure 8). The image name is locked (i.e., you can not edit it) so that the file attachment in the desktop library will remain intact. Actual images can not be added to the EndNote application on PDA. If you

Sample EndNote Library.enl

	Author	Year	Title
	CBE St...	1994	Scientific style and format : the
	Health ...	2001	Back to basics : foundations c
	Iverson	1998	American Medical Associatio
	Jenny-A...	2002	Obesity and the risk of heart fa
	Kendrick	2001	Radiography of the lumbar spi
	Kohn	2000	To err is human : building a sa
	Kuperm...	2003	Computer physician order entr
	Levinson	2000	Diagnosis and treatment of re

Showing 99 out of 99 references. Show Preview

FIGURE 6. Selecting references in EndNote library

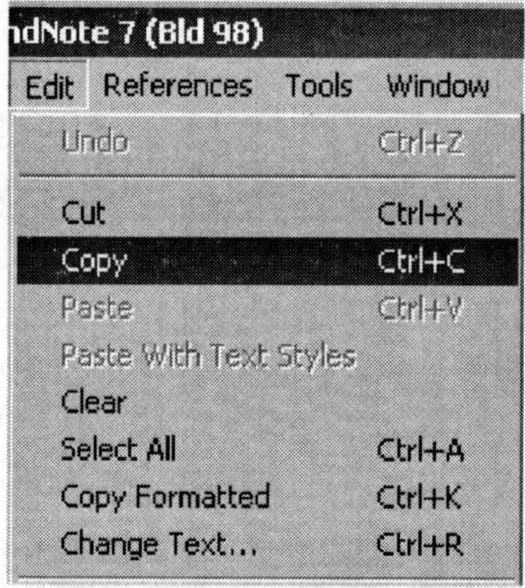

FIGURE 7. Selecting Edit > Copy

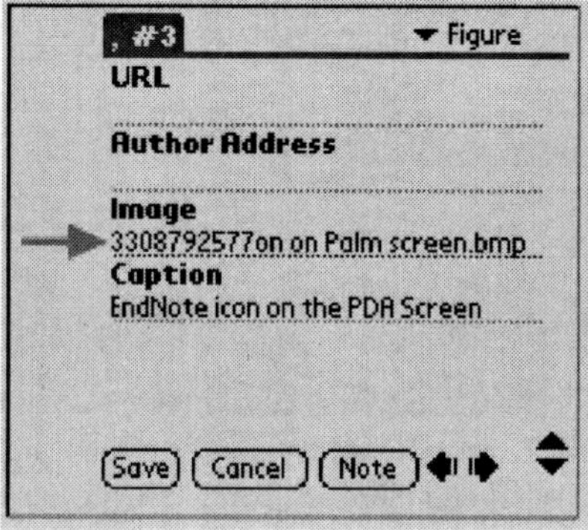

FIGURE 8. Figure type reference in PDA

FIGURE 9. Opening the EndNote Library on PDA

delete a reference with an image from your PDA, synchronizing will delete both the reference and its corresponding image from the desktop library.

Working with the EndNote Library on PDA

Opening the Library

- From the applications screen of your PDA, tap on the EndNote icon (Figure 3).
- After the splash screen disappears, you will see the name of the EndNote library (Figure 9). Highlight the name of the library and tap *Open*.
- A reference list appears displaying three columns—Author (the last name of the first author), Year, and Title (Figure 10). Note that the title bar shows the name of the current library and the author name, year, and the reference title are truncated to fit the screen of the PDA.

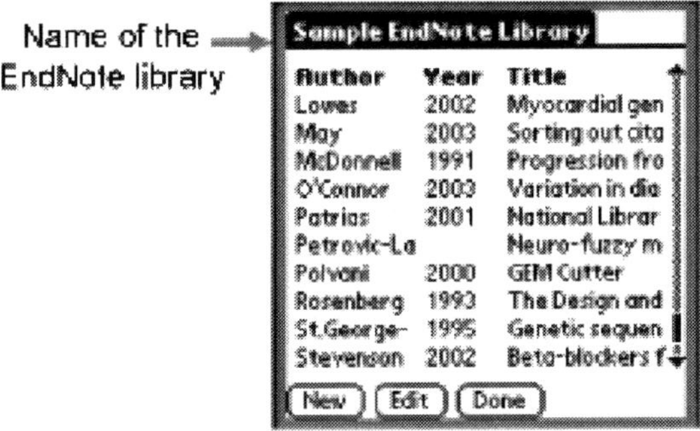

FIGURE 10. Displaying references on the PDA screen

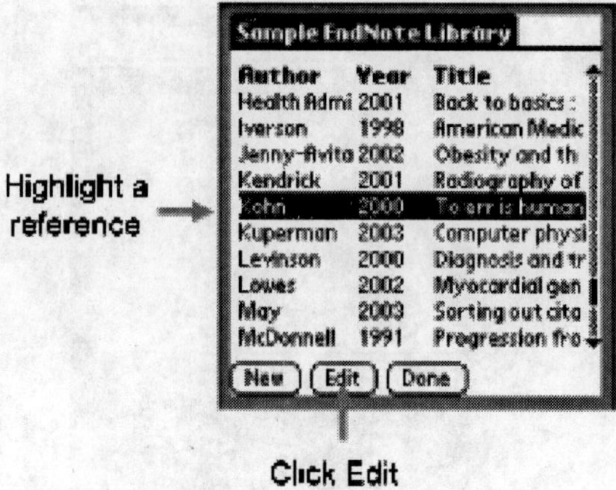

FIGURE 11. Opening a reference in the PDA library

Opening a Reference for Viewing or Editing

- Tap twice on the row of that reference.

  OR

- Highlight the reference and tap *Edit* (Figure 11).

Sorting References

- Tap on a column header to sort the list by that column of information. For example, if you tapped the column header "Author," it will arrange your references alphabetically by the last name of the first author.
- Tap the same column header again to switch between ascending and descending order of sorting.

Similarly, tapping on the "Year" header will sort references by the year and tapping on the "Title" header will sort the references alphabetically by the title.

Entering a New Reference

- Open an EndNote library and display the reference list.
- Tap *New* to display a New Reference template (Figure 12).
- Select a reference type by tapping the dropdown list (Figure 13). This dropdown list is similar to the EndNote library on the desktop. The EndNote for Palm supports all of the reference types and fields used in the desktop EndNote application.
- Enter text into appropriate fields using the keyboard or Graffiti™, the text writing tool of the PDA.
- Tap *Save* to add the reference to library.

Now synchronize your PDA with the desktop using HotSync to see the newly added reference in the EndNote library on the desktop.

FIGURE 12. Entering a new reference in PDA

Editing a Reference

There are two ways to edit a reference in your EndNote library

First Method (Figure 11)

- Open an EndNote library and display the reference list.
- Highlight the reference you want to edit by tapping the reference.
- Select *Edit* from the reference display window.

Second Method (Figure 14)

- Open an EndNote library and display the reference list.
- Highlight the reference you want to edit by tapping the reference.
- Tap *Menu* button.
- Select *Edit* from the dropdown list.

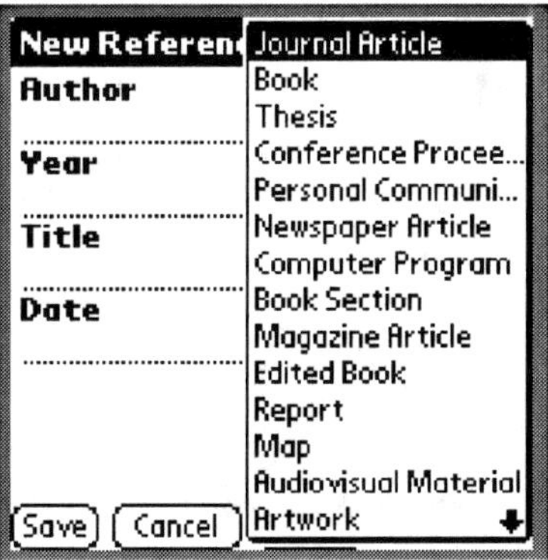

FIGURE 13. Selecting a reference type

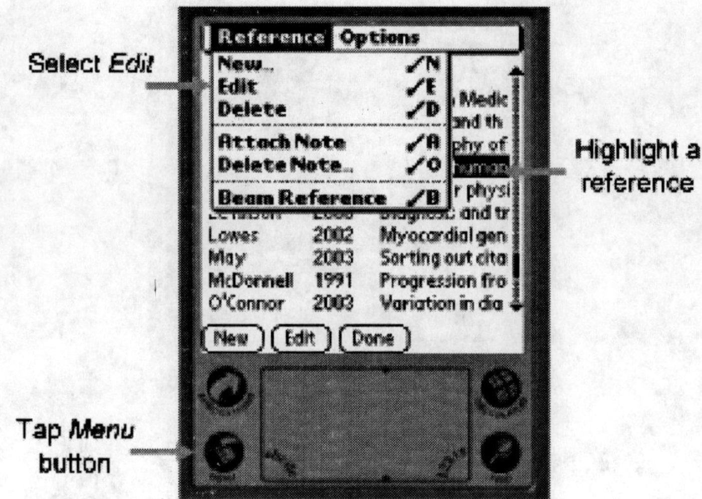

FIGURE 14. Editing a reference by using the Menu button

Deleting a Reference (Figure 15)

- Open an EndNote library and display the reference list.
- Highlight the reference you want to delete by tapping the reference.
- Tap *Menu* button.
- Select *Delete* from the dropdown list.

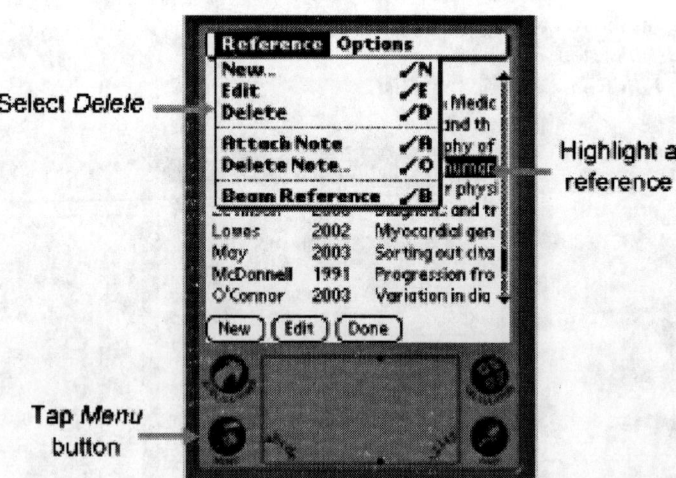

FIGURE 15. Deleting a reference

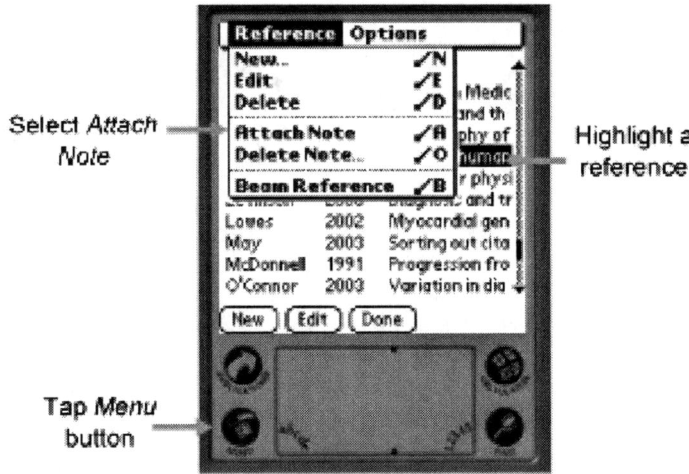

FIGURE 16. Attaching a note to a reference

Attaching a Note to a Reference (Figure 16)

- Open an EndNote library and display the reference list.
- Highlight the reference you want to attach a note to by tapping the reference.
- Tap *Menu* button.
- Select *Attach Note* from the dropdown list. To delete a note, select *Delete Note* instead.

Searching References

If you are looking for a particular reference, you can jump to it in the reference list or you can search by keywords by using the *Find* icon.

To jump to a particular reference (Figure 17)

- Sort the library by a column, for instance, by author or by title.
- Enter characters in the text entry area to scroll to that text in the library. For example, if the current sorting is by author, entering "ste" will scroll the list to Stevenson. You would notice that now a *Look Up field* (with the text "ste") has appeared on the screen (Figure 18). You can modify your search by changing the text in this field.

To Search by Keyword

- Open an EndNote library and display the reference list.
- Tap on the *Find* icon (looks like a magnifying glass) on the PDA (Figure 19).
- Enter the keyword you wish to search.
- Tap *OK.*

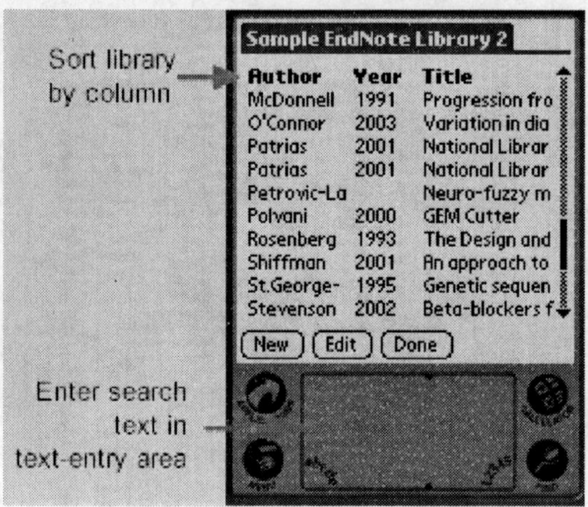

FIGURE 17. Searching references by jumping to a particular reference

FIGURE 18. Searching references using the Look-up field

FIGURE 19. Searching references using the Find icon

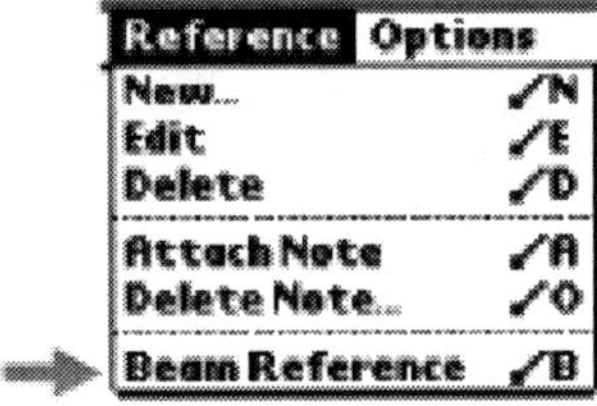

FIGURE 20. Beaming references from PDA

Beaming References Between PDAs

'Beaming' selected references to another PDA using the infrared port is similar to beaming your business card or other material from one PDA to the other. The receiving PDA must also have the EndNote for Palm OS application installed for the beaming to work.

> ⚠ **Alert:** You cannot beam the entire EndNote library or the EndNote for Palm application itself. Also, both the sending and the receiving PDA must have the Palm OS. You cannot beam EndNote application or references to a PocketPC based PDA.

Beaming a Reference or a Group of References

- Open an EndNote library in your PDA, display the list of references.
- Highlight a reference or a group of references by tapping on the reference rows.
- Tap *Menu* button. From the *Reference* menu, tap *Beam Reference* (Figure 20). You should see the following message on your PDA (Figure 21).
- On the receiving PDA, tap *Yes* on the *Beam Receive dialog* to accept the reference(s).

How Beaming Works

All modern PDAs come equipped with an infrared beaming port which allows PDA users to share data wirelessly between two PDAs within one meter

FIGURE 21. Beaming message on PDA

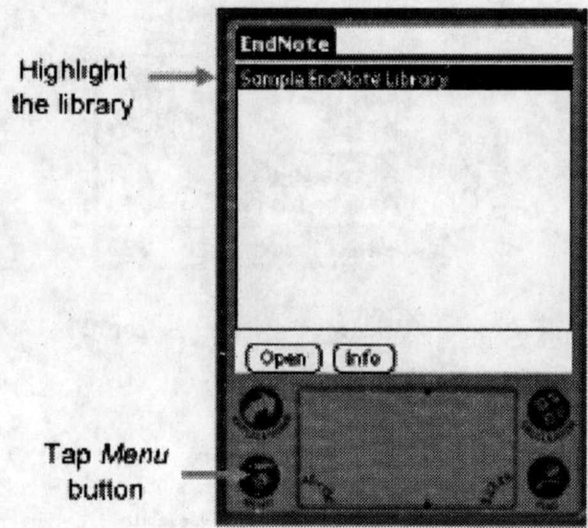

FIGURE 22. Viewing statistics: Getting to the library menu

(39.37 inches). It works similar to how a TV remote control beams instructions to television.

Viewing Statistics About the PDA EndNote Library

EndNote provides you the facility to look up basic information about the PDA EndNote library, such as the date on which the library was created and last modified, the number of references in the library, and the memory this library is occupying on the PDA.

To view this information:

- Start EndNote for Palm by tapping on the EndNote icon.
- Highlight the name of the library.
- Tap the *Menu* button (Figure 22).
- Tap *Library* menu.
- Tap *Info* to display statistics about the highlighted EndNote library (Figure 23).
- You will see the information about your EndNote library in the next screen (Figure 24).
- Tap the *OK* button to exit out of the information display.

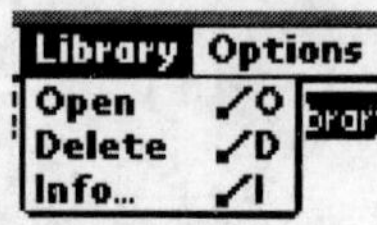

FIGURE 23. Selecting Info from the library menu

FIGURE 24. Displaying library information

Customizing the PDA EndNote Library

- Start EndNote for Palm by tapping on the EndNote icon.
- Highlight the name of your library.
- Tap the *Menu* button (Figure 22).
- Tap *Options* menu.
- Tap *Preferences* to set preferences for the highlighted library (Figure 25).

You can select from the following options (Figure 26):

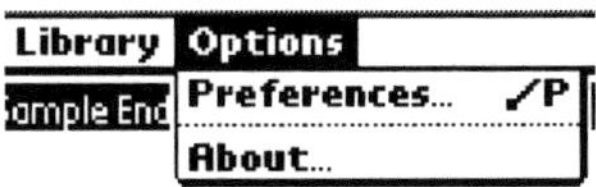

FIGURE 25. Selecting Preferences menu

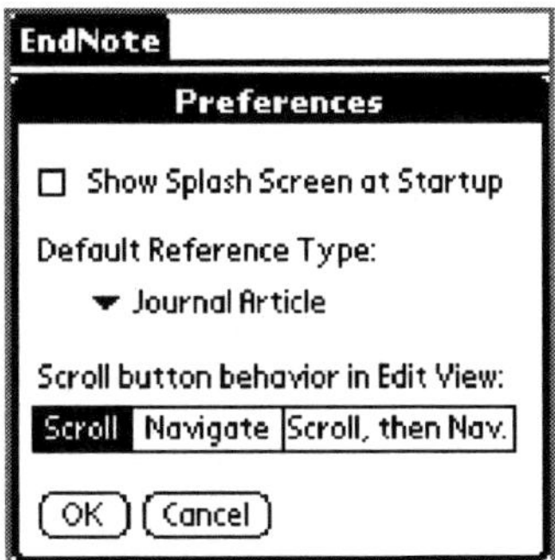

FIGURE 26. Customizing EndNote on PDA

FIGURE 27. EndNote splash screen

Show Splash Screen at Startup

Generally when you open the EndNote application on your PDA, a brief splash screen appears with information about the EndNote application (Figure 27). Use the check-box option to enable or disable the splash screen display (Figure 26).

Default Reference Type

Use this list to select a default reference type to assign to new references added to your PDA. If most of the time you are entering a particular type of reference, such as a journal article, or a book, then it would save you time by not having to select a reference type every time you enter a new reference. Remember, you can always change the reference type when actually entering a reference.

Scroll Button Behavior in Edit View

This allows you to select the behavior of scrolling buttons while in Edit view.

- *Scroll*: The scroll button scrolls through the fields of current references
- *Navigate*: The scroll button navigates forward and backward through references
- *Scroll then Nav.*: the scroll button scrolls until the end of the current reference, then navigates to the next reference.

11
Using RefViz© with EndNote

> ## Things You will Learn in This Chapter
>
> - What RefViz is.
> - How RefViz works.
> - Technical requirements for installing RefViz.
> - How to get, install, and activate RefViz.
> - How to work in various RefViz views, including the Galaxy and the Matrix views.
> - How to export references from RefViz back to EndNote.

This chapter provides you with a basic introduction to RefViz, a data visualization software program. The RefViz program is completely different from, and independent of, EndNote. You can use EndNote without RefViz. In addition to EndNote, RefViz is useful to analyzing data from a variety of sources, such as ProCite and Reference Manager.

This chapter will help you understand the concept of RefViz as a data visualization program and to decide whether you want to buy or try this program further. A more detailed discussion of RefViz is beyond the scope of this book. If you do decide to buy or try RefViz, it comes with a user-friendly Help file which will help you work with this program.

What is RefViz?

EndNote is a great application for storing, managing, searching, and citing references in a bibliography. If you want to further analyze this collection of references, data visualization programs such as RefViz provide enhanced data mining capability.

RefViz (www.refviz.com) works as a companion application for EndNote to enhance EndNote's functionality. It works seamlessly with EndNote in that it can be launched from within EndNote and selected groups of references from RefViz can be exported back into an EndNote library.

How Does RefViz Work?

RefViz processes selected references from EndNote library and does the following:

- It analyzes the information present in the Title and Abstract fields for each reference. If the Abstract field is empty, it will use the Notes reference field instead. If both these fields are empty, the reference is normally removed from the RefViz analysis (although you can set the preferences in RefViz to analyze data in these references anyway).
- It applies mathematical algorithms to this information to determine the key concepts in a library and automatically generates a list of keywords. Note that these keywords are different from the information in the Keywords reference fields in the EndNote library.
- It organizes and groups your references along these keywords and produces an interactive visual display in two formats: the Galaxy View and the Matrix View (see details about these two in next sections).

Below is a flowchart of RefViz processes.

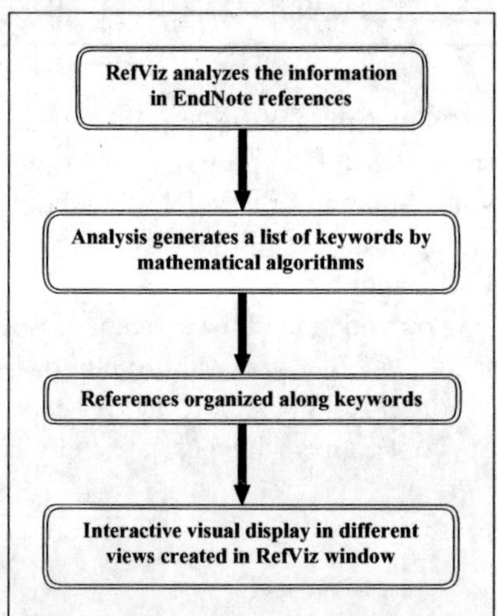

Technical Tip: RefViz does *not* use predefined categories such as MeSH Terms or Keywords from EndNote reference fields to create reference groups. It uses its own statistical analysis of the text in the Title and Abstract/Notes field to create keywords for a particular analysis.

Technical Requirements for Installing RefViz

RefViz works under both Windows and Mac operating systems. The following are the requirements to use RefViz on Windows computers:

- Windows NT / 2000 / ME / XP.
- Pentium 3 or higher, 1 GHz recommended (200 MHz minimum).
- 256 MB RAM recommended (128 MB minimum).
- 1 GB free disk space (100 MB minimum).
- 1024 × 768 monitor resolution (800 × 600 minimum).
- RefViz requires JAVA 1.4.1 or later.

Getting and Installing RefViz

You can buy RefViz from the web site of ISIResearchSoft at www.refviz.com. You can also download a free trial version from the above web site.

> ⚠ **Alert:** Note that if you use a version of EndNote prior to version 7, you can still use RefViz but this will involve an intermediate step. You will need to export your EndNote library as a text file in RIS format and then import that file into RefViz.

Installing RefViz

- Double-click on the RefViz installation file. As usual, close all applications before proceeding with the installation.
- Simply follow the instruction on the next screens and RefViz will be installed on your computer.
- To check that RefViz has been installed successfully, click *Start* > *Programs* > *RefViz 1.0* > *RefViz* (Figure 1). This should launch the program.

Activating RefViz

You can launch RefViz from EndNote screen by using EndNote's Data Visualization command. However, you need to 'activate' this command. Without activation, this command appears grayed out in EndNote (Figure 2). The activation needs to be done only once. To perform activation:

- Locate the EndNote folder on your computer. If you accepted the default option during EndNote installation, the most likely path for this folder is C:\ProgramFiles\EndNote.
- In this folder, double click on the file named "Configure Omniviz.exe."

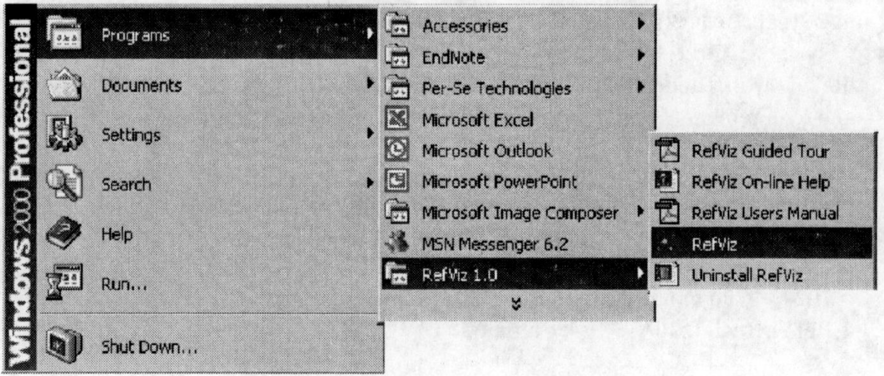

FIGURE 1. Checking RefViz installation

- Follow the instructions on the screen. In the second screen, you will be asked to select the location of the RefViz application (Figure 3). **You must select the correct location for the RefViz file; otherwise the activation will not work.**
  - Click on the *Browse* button.
  - If you accepted default options during installation, the path to the RefViz folder will be C:\Program Files \ RefViz.
  - Select the RefViz.exe file. Click Open (Figure 4).
  - Continue with the installation.

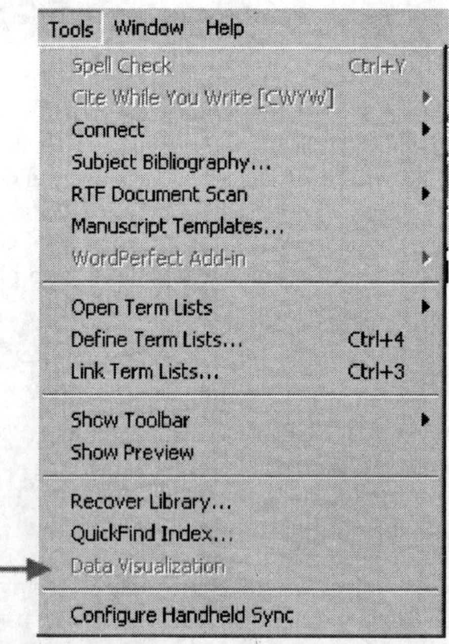

FIGURE 2. Inactive Data Visualization command in EndNote

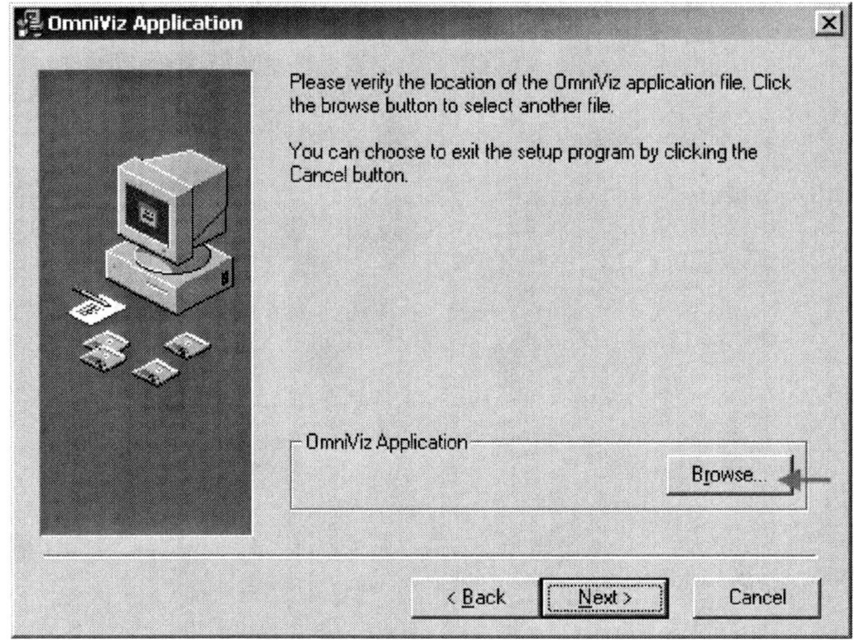

FIGURE 3. Screen to verify the location of RefViz application

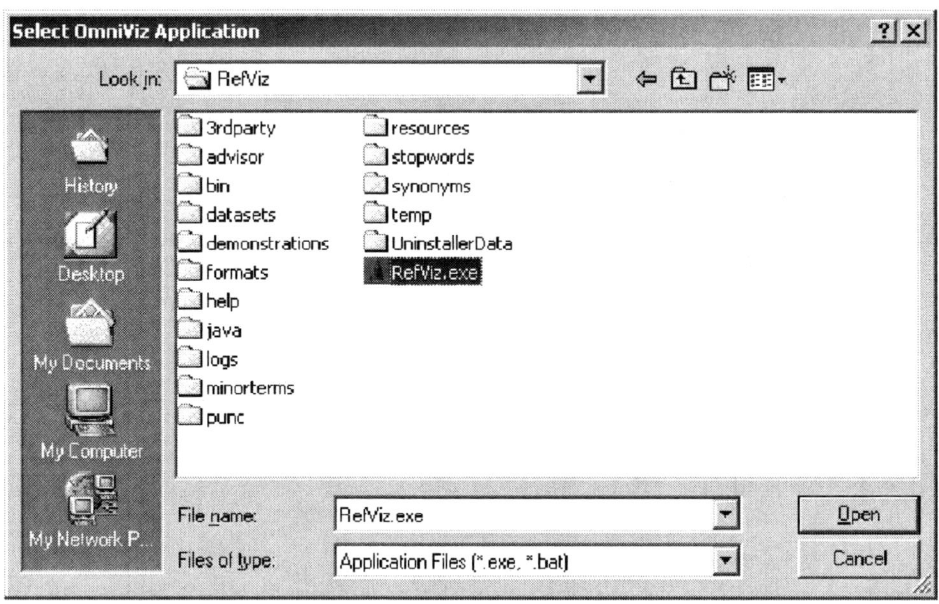

FIGURE 4. Selecting the RefViz.exe file

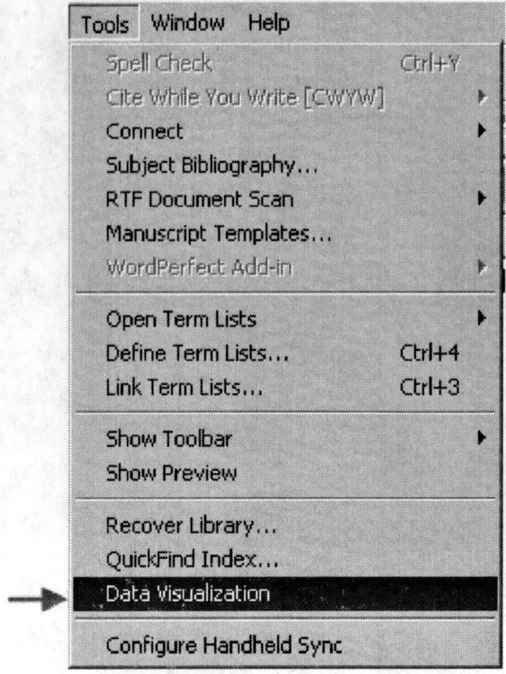

FIGURE 5. Activated Data Visualization command in EndNote

- The Data Visualization command should now be active in EndNote now (Figure 5).

Technical Tip: If EndNote was open during the activation, close it and restart the EndNote program. If this procedure doesn't activate the Data Visualization command, you may need to download an output style from the web site of RefViz at http://www.refviz.com/support/rvimport_ris.asp (Figure 6).

Working in RefViz

Starting RefViz

Click *Tools > Data Visualization* from EndNote (Figure 5). RefViz will launch automatically, import references from EndNote, and organize them into the RefViz window for display.

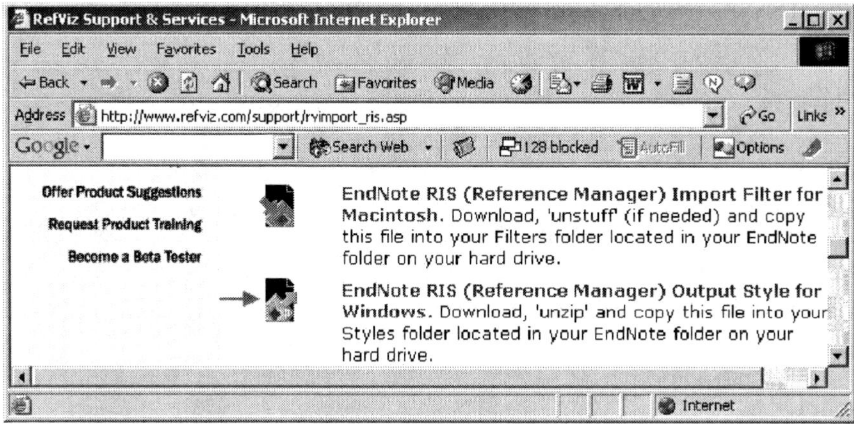

FIGURE 6. Downloading the EndNote RIS output style from the website of RefViz

Working in RefViz

The following is a general overview of various parts of the RefViz window and the functionality provided in these parts. Note that I have used the sample references provided in the trial version of RefViz in various illustrations.

The RefViz window is divided in four main parts, with each part containing a different view and function for reference analysis.

Galaxy View (Figure 7)

- Located in the upper left corner. By default the galaxy view tab is selected.
- In this view, each reference is represented by a small square.
- Groups of related references are represented by a paper icon such as .
  - The larger the paper icon, the higher the number of references in that group.
  - Related groups of references (i.e., paper icons) are located near each other. So this view provides a proximity map of references as well.
- Hover your mouse over any paper icon—it will give you a summary of information about references in that group. In Figure 7, the arrow indicates the paper icon on which the mouse was hovering at the time of taking this screen shot.
- If you click on a paper icon, that icon remains 'highlighted'; all others icons are darkened (Figure 8). In addition, individual references in that group appear in the reference viewer window.
- Use the pan, zoom, and magnify icons in the toolbar to focus on references in a particular group in the galaxy (Figure 7).

Matrix View (Figure 9)

- Located in the upper left corner. Click on the matrix tab to get to this view.
- This view provides a visual display of relationship between keywords and groups.

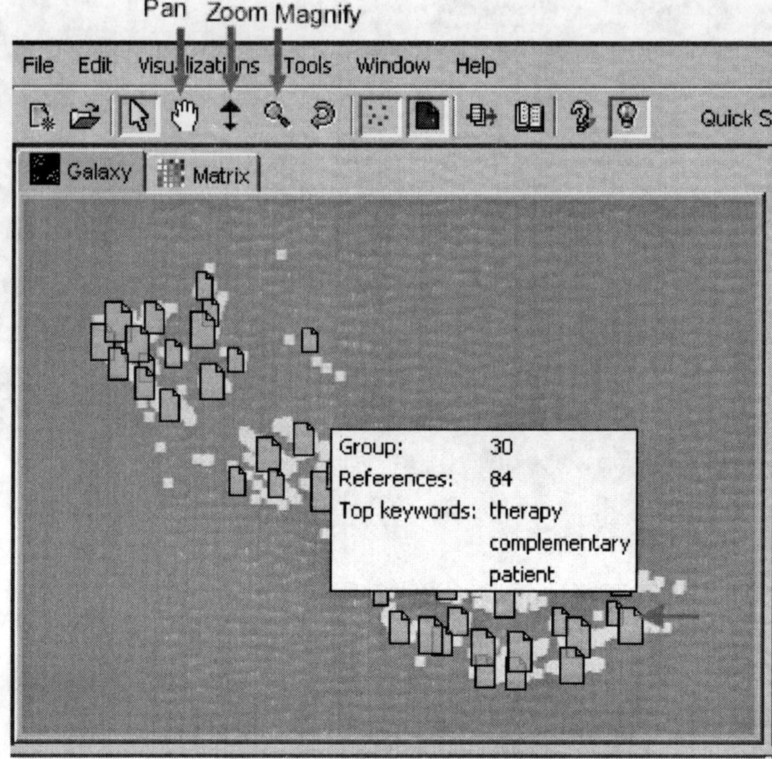

FIGURE 7. The Galaxy view in RefViz

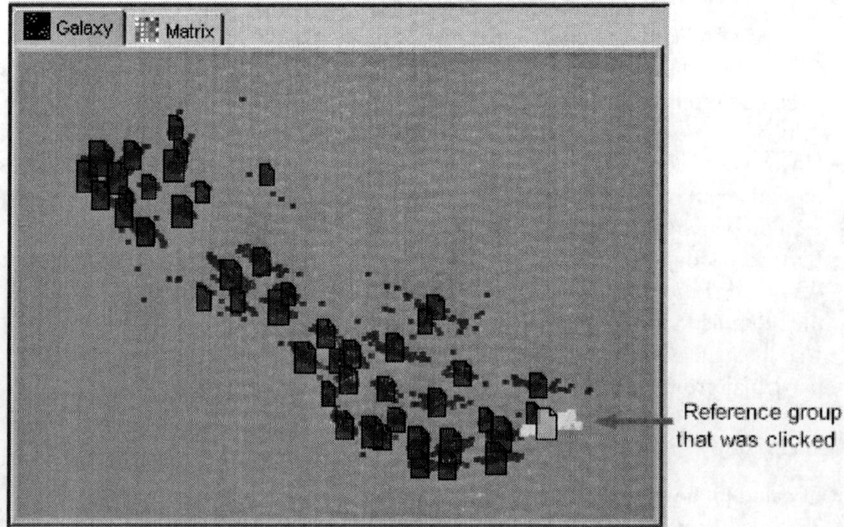

FIGURE 8. The Galaxy view after a reference group in clicked

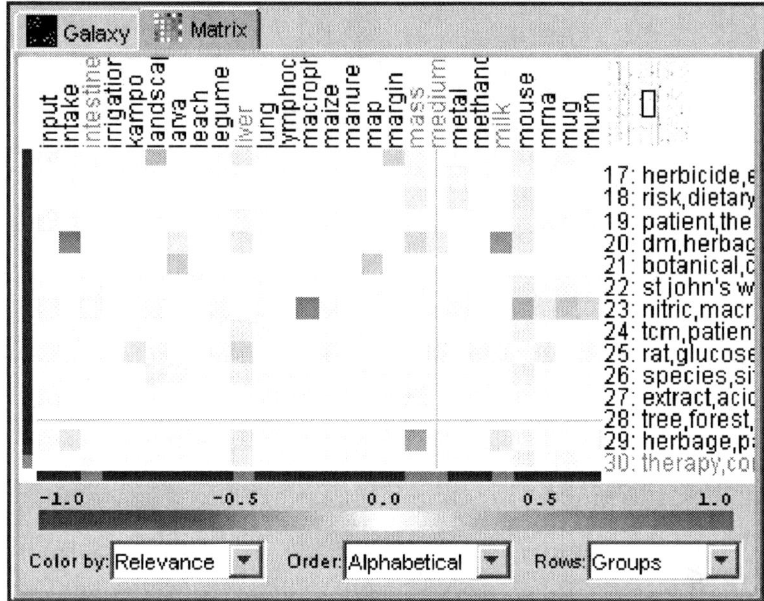

FIGURE 9. The Matrix view in RefViz

- By default, keywords are displayed in columns and reference groups are displayed in rows.
- The color of the cells of the matrix indicate the strength of the relationship, that is, how often the corresponding terms in keywords and groups are discussed together:
  - Red cells indicate close relationship.
  - Blue cells indicate weak relationship.
  - White cells indicate no significant association.

Reference Viewer Window (Figure 10)

- Located in the lower left corner.
- Displays information about selected references. Each reference is in one row and each reference field is in a column (similar to the display pattern in the EndNote library window).
- Double—clicking on a reference will display it in a separate pop-up window for easy viewing of references (Figure 11).
- You may choose the fields to be displayed in this window by clicking *Edit > Reference Viewer > Choose Fields* (Figure 12). Select column fields from the list in the next window and click *OK* (Figure 13).

Reference Viewer

Group	Title	Authors	Pub Date
3	Effects of organic fert...	Saitoh, K; Haya...	2002
3	Biomass production a...	Hauser, S; Nolte,...	2002
3	Impacts of short-rotati...	Perry, CH; Miller,...	2001
3	Sediment and nutrient ...	Craft, CB; Casey...	2000
3	Asymmetrical distributi...	Dowdy, RH; Dol...	2000
3	Occurrence of Phytop...	Aldaoud, R; Gup...	2001
3	The current and poten...	Kennedy, IR; Isla...	2001
3	Organic amendments t...	Moorman, TB; C...	2001
3	Forage legumes for im...	Muhr, L; Tarawa...	1999
3	Comparison of energy...	McLaughlin, NB; ...	2000
3	Atrazine and nitrate-ni...	Chinkuyu, AJ; K...	1999
3	Medieval sheep-corn f...	Newman, EI	2002
3	Soil fauna, guilds, fun...	Brussaard, L	1998
3	Response of sorption ...	Haberhauer, G; ...	2001
3	Genotoxicity of conta...	Kong, MS; Ma, TH	1999
3	Cs-137 and Sr-90 mo...	Korobova, E; Er...	1998

FIGURE 10. The Reference Viewer window

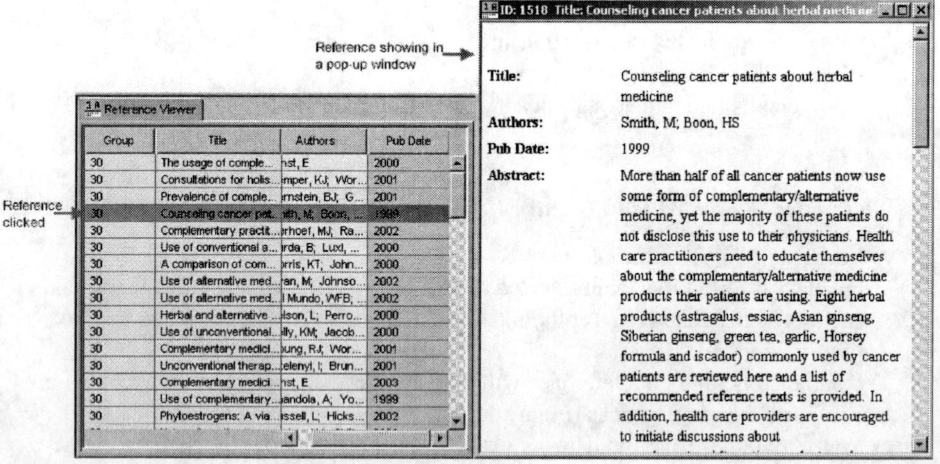

FIGURE 11. Clicking on a reference in the Reference Viewer window

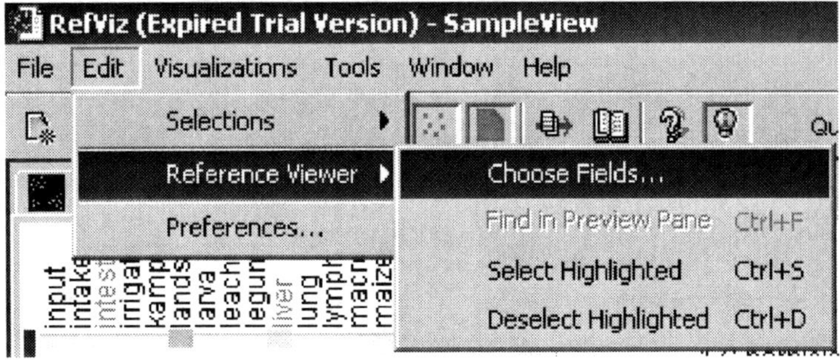

FIGURE 12. Selecting Reference Viewer > Choose Fields

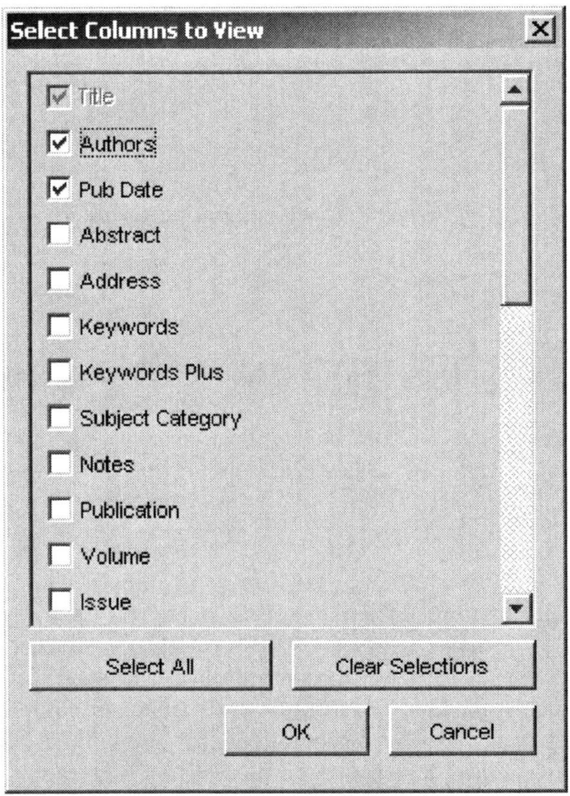

FIGURE 13. Selecting columns fields for display in the Reference Viewer

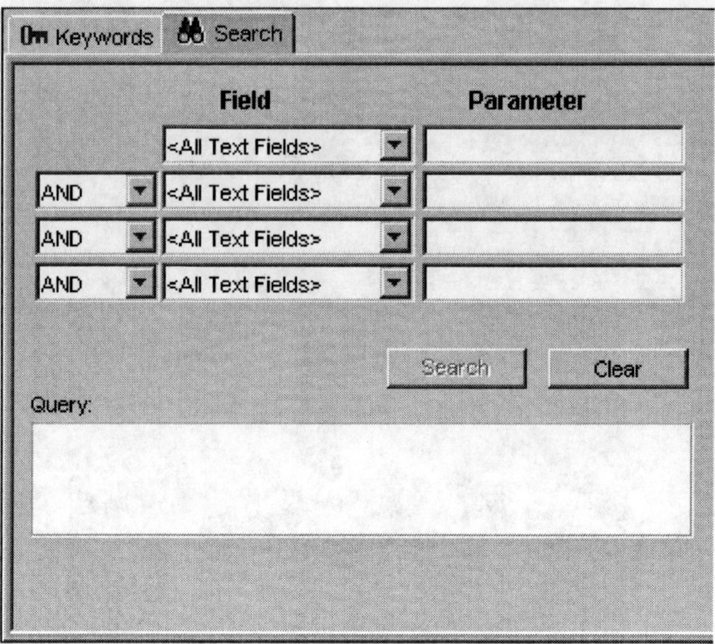

FIGURE 14. The Search tool in RefViz

Search Tool (Figure 14)

- Located in the upper right corner.
- Use the search tool to perform a search of references by content in various reference fields.
- The references yielded by the search will automatically populate the reference viewer as well as various visualizations.

Keyword Tool (Figure 15)

- Also located in the upper right corner.
- Displays the vocabulary of the reference set and how it was categorized during processing.
- A useful tool to explore themes in your references and groups.
- Primary keywords are the terms that best distinguish sets of references
- Secondary keywords are less distinguishing, but also have an influence on the grouping through their co-occurrence with the primary keywords.
- Other Descriptive Terms make up the rest of the vocabulary in the titles and abstracts of your references. They are either too frequent or too infrequent to help define groups of references, or they are not distributed in a manner that would help distinguish groups.

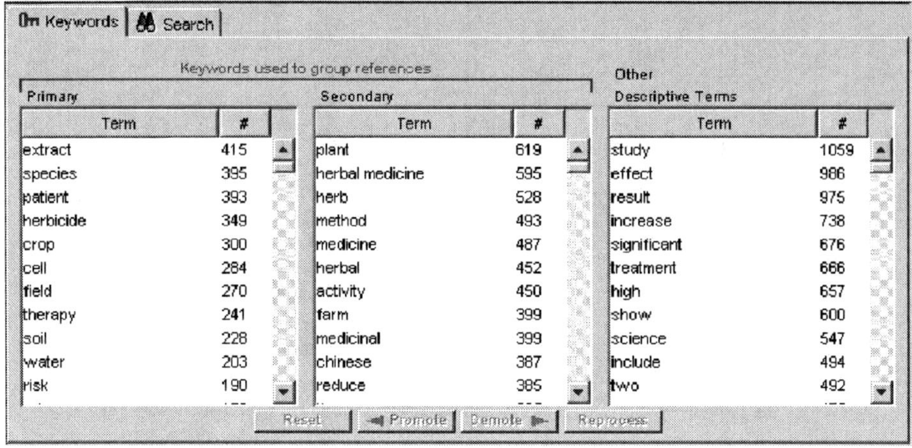

FIGURE 15. The Keyword tool in RefViz

Advisor (Figure 16)

- Located in the lower right corner.
- Provides easily available context-dependent information about various RefViz views.
- The contents of the advisor are related to the view you are working on in RefViz. For example, if you are working in the galaxy view the advisor will display the information about the galaxy view and if you clicked on the matrix view, the information will automatically change to the matrix view.

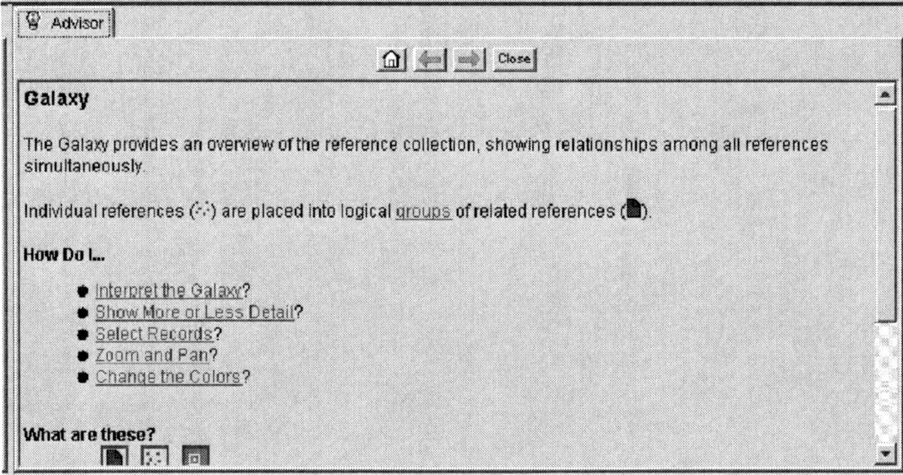

FIGURE 16. The Advisor in RefViz

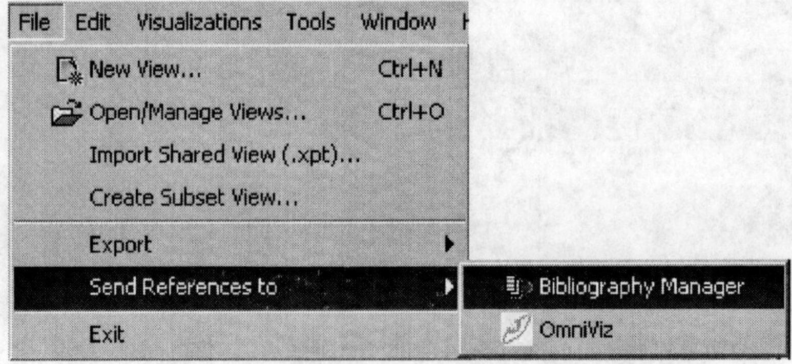

FIGURE 17. Exporting selected references from RefViz to EndNote

Exporting References from RefViz Back to EndNote

After you have analyzed references in RefViz, you may want to export a selected set of references back into an EndNote library. To do this, first decide which EndNote library you wish to export these references to or create a new blank EndNote library.

- Identify and select references in RefViz using one of the windows such as Reference Viewer, Search tool, or groups of references in the Galaxy or Matrix visualizations.
- Click *File > Send References to > Bibliography Manager* in RefViz (Figure 17).
- In the next screen, browse and select an EndNote library. References will be sent from RefViz to EndNote.

Appendix A
Online Resources to Learn More About EndNote

- **EndNote website**
  http://www.endnote.com
- **EndNote online technical support**
  http://www.endnote.com/support/ensupport.asp
- **Australian Catholic University Library**
  http://dlibrary.acu.edu.au/endnote/entutorials.htm
- **Curtin University of Technology Library**
  http://library.curtin.edu.au/referencing/endnote.html
- **Duke University Medical Center Library**
  http://www.mclibrary.duke.edu/training/endnote
- **Harvey Cushing/John Hay Whitney Medical Library, Yale School of Medicine**
  http://www.med.yale.edu/library/reference/training/endnote/welcome.html
- **Leeds University Library**
  http://www.leeds.ac.uk/library/training/endnote/
- **Monash University Library**
  http://www.lib.monash.edu.au/vl/endnote/endncon.htm
- **RMIT University Library**
  http://www.rmit.edu.au/library/endnote
- **The Australian National University**
  http://ilp.anu.edu.au/endnote/
- **The University of Queensland Library**
  http://www.library.uq.edu.au/faqs/endnote/faqs.html
- **University of Michigan's Knowledge Navigation Center**
  http://www.lib.umich.edu/knc/howto/citation/endnote.html#intro
- **University of North Carolina Health Sciences Library**
  http://www.hsl.unc.edu/services/tutorials/endnote/intro.htm
- **Yale University Library**
  http://www.library.yale.edu/endnote/

Appendix B
Online Resources to Help Writing for Publication

Finding Authors' Guidelines

- **International Committee of Medical Journal Editors (ICJME)**
  http://www.icmje.org/index.html#manuscript
- **Instructions to Authors in the Health Sciences, Medical College of Ohio**
  http://www.mco.edu/lib/instr/libinsta.html

Finding Journal Title Abbreviation

- **Journal Abbreviation Sources (JAS)**
  http://www.public.iastate.edu/%7ECYBERSTACKS/JAS.htm
- **Locator Plus by the National Library of Medicine**
  http://locatorplus.gov/

Citing Internet Sources

- **American Psychological Association (APA)**
  http://www.apastyle.org/elecref.html
- **Beyond the MLA Handbook: Documenting Electronic Sources on the Internet**
  http://english.ttu.edu/kairos/1.2/inbox/mla_archive.html#varma
- **Columbia Guide to Online Style**
  http://www.columbia.edu/cu/cup/cgos/idx_basic.html
- **International Organization for Standardization (ISO 690-2)**
  http://www.collectionscanada.ca/iso/tc46sc9/standard/690-2e.htm
- **Modern Language Association (MLA)**
  http://www.mla.org/bib_guidelines_elec
- **National Library of Medicine**
  http://www.nlm.nih.gov/pubs/formats/internet.pdf

- **Purdue University Online Writing Lab**
  http://owl.english.purdue.edu/handouts/research/r_docelectric.html
- **The Library of Congress: How to Cite Electronic Sources**
  http://lcweb2.loc.gov/ammem/ndlpedu/start/cite/index.html
- **University of Alberta**
  http://www.library.ualberta.ca/guides/citation/index.cfm#biomedical

Appendix C
Downloading Files, Filters, and Styles for EndNote

Downloading Connection Files and Filters

- **EndNote website**
  http://www.endnote.com/support/ensupport.asp
- **University of Canberra, Australia**
  http://www.canberra.edu.au/library/research-gateway/research_help/endnote/filter-files
  http://www.canberra.edu.au/library/research-gateway/research_help/endnote/connect-files
- **University of Technology, Sydney**
  http://www.lib.uts.edu.au/information/endnote/filters.html
- **The Australian National University**
  http://ilp.anu.edu.au/endnote/how_filters.html
- **The University of New South Wales**
  http://www.library.unsw.edu.au/%7Epsl/helpsheets/endnote_filters.html

Downloading Output Styles

- **EndNote website**
  http://www.endnote.com/support/enstyles.asp
- **Auckland University of Technology**
  http://www.isworld.org/endnote/endnote.htm
- **The University of Georgia Libraries**
  http://www.libs.uga.edu/liaison/endnote/downloadstyles.html
- **The University of New England**
  http://www.une.edu.au/library/endnote/endnote_styles.html
- **The University of Queensland**
  http://www.library.uq.edu.au/faqs/endnote/styles.html#alpha
- **The University of Western Australia Library**
  http://www.library.uwa.edu.au/guides/endnote/styles.html

Index

About EndNote menu, 22
About EndNote screen, 22*f*
Abstracts, online
  citing references from Internet sources, 183
Add-in Support Option, 25*f*
APA 5th style, 137, 138*f*
Articles, online
  citing references from Internet sources
    journal articles, 183
    magazines, 183
    newspaper, 183
  downloading references from Websites of Journals, 75*f*, 76*f*
Attaching note to a reference, 205
Authors
  Internet, citing references from sources on, 178
"Auto-completion," 66
Automatically updating, 33–34

Backup
  folder, 20
  selecting backup of files, 21*f*
Beaming references between PDAs, 207–208
Bibliographies
  bookmark, creating, 170–171
  chart references in manuscripts, working with, 149–152
  cited references, finding and editing, 146
  Cite While You Write (CWYW)
    output styles, 132–133
    preferences, 168–169
  creating, 144–149
    cited references, finding and editing, 146
    customizing, 144–146
    font, customizing, 145–146
    Format Bibliography, 145*f*
    formatting, 144
    Generate Figure List command, 146

Bibliographies (*cont.*)
  creating (*cont.*)
    layout, customizing, 145–146
    manuscript, 139–140
    multiple documents, from, 146–148
    notes, inserting in list of references, 148–149
    Outline View, selecting, 147*f*
    placement, customizing, 146
    subdocuments, inserting, 147*f*
  cross references, creating, 171–173
  customizing, 144–146
    figures and tables/charts, 151–152, 153*f*
  field shading, turning off, 172*f*, 173
  figure references in manuscripts, working with, 149–152, 153*f*
  Find Figure(s) dialog, 151*f*
  font, customizing, 145–146
  footnotes, citing references in, 157–159
    customizing citations, 159–160
  Format Bibliography, 145*f*
  formatting, 144
  Generate Figure List command, 146
  independent bibliography
    Copy Formatted command, 161*f*
    copy-paste method, 161
    creating, 160–162
    defined, 160
    drag-and-drop method, 161
    export method, 162
    print method, 162
  layout, customizing, 145–146
  manuscript, creating, 139–140
  manuscript, inserting references from EndNote library into, 140–144
    changing existing citations, 142–144
    copying citations, 143
    CWYW menu, 141*f*

Bibliographies (*cont.*)
manuscript, inserting references from (*cont.*)
deleting citations, 144
editing citations, 142
Find Citations Dialog, 141*f*
moving citations, 143
setting EndNote formatting preferences, 143*f*
unformatting citations, 142–143
multiple documents, from, 146–148
notes, inserting in list of references, 148–149
Outline View, selecting, 147*f*
output styles, 132–138
APA 5th style, 137, 138*f*
authors' guidelines for biomedical journals, 133*t*
Cite While You Write (CWYW), 132–133
Edit Citations dialog, 138*f*
editing styles, 135–138
examples of editing styles, 137–138
favorites, marking styles as, 135
Format Bibliography command, 136*f*
Style Editor window, 136*f*
style manager, 134–138
overview of steps, 131–132
page number, inserting, 170
placement, customizing, 146
publisher, sending paper to, 152–157
field codes, 152–154
field codes hidden, citation with, 154*f*
field codes showing, citation with, 155*f*
generally, 157
removing field codes, 157
Toggle Field Codes, 155*f*
traveling library, 155–156
reference list and, 2–3
reference management software program, 9
sharing document with others, 155–156
subdocuments, inserting, 147*f*
subject bibliography, creating, 163–168
customizing, 165, 166*f*
definition of subject bibliography, 163
generally, 163–165
printing, 165–166
sample subject bibliography, 165*f*
saving, 165–166
selecting subject fields, 164*f*
selecting subject terms, 164*f*
subject list, creating, 163–168
definition of subject list, 163
generally, 166–168
table/chart references in manuscripts, working with, 149–152
Biomedical databases, commonly used, 102*t*
Blank electronic source box, 191*f*
Book chapters, online
citing references from Internet sources, 184
Bookmark, creating, 170–171
Books, online
citing references from Internet sources, 183
Boolean operators, 87*t*

Change fields, 94, 95*f*
Change Text command, 92–93
Character Map program, 73*f*
Chart references in manuscripts, working with, 149–152
Chart type references, 68–72
Checking installation
Add-in Support Option, 25*f*
Custom Installation, selecting, 24*f*
CWYW files, 25*f*
File Locations, Startup folder, 27*f*
Help, 34*f*
"Hide file extensions for known file types," unchecking, 29*f*
Macro Security dialog box, 34*f*
Microsoft Word, 23, 24*f*, 25
Macro Security dialog box, 34*f*
Macro Security Warning, 31, 32*f*
Security Level tab, Medium setting, 32*f*
shortcuts not working, 31
troubleshooting, 25–31
Security Level tab, Medium setting, 32*f*
"Show hidden files and folders," selecting, 30*f*
Startup folder, 27*f*, 28*f*
troubleshooting, 25–31
Trust All . . . box, 33*f*
Windows Explorer
folder options, selecting, 29*f*
launching, 28*f*
word processor support, checking, 23–33
Citations. *See specific topic*
Cite While You Write (CWYW), 25*f*, 132–133, 141*f*
figures and tables, 154*f*
preferences, bibliographies, 168–169
preferences dialog box, 168*f*
Comparison operators, 86*t*
Connection file method, 106–114
choice of, 104–105
defined, 106
described, 102
downloading connection files from Internet, 107–109
empty retrieved references window, 111*f*

Connection file method (*cont.*)
establishing connection to database, 110
"favorite" connection files, setting, 107
illustrated, 108*f*
number of references found by search, window displaying, 113*f*
opening Connection Manager, 106*f*
PubMed, 118–119
selecting, 110*f*
replacing old connection file warning, 109*f*
retrieving references, 111–114
saving connection file to computer, 109*f*
saving references to EndNote Library, 114
searching database, 111
search window with search terms, 112*f*
selecting connection file for download, 108*f*
using method, 109–114
working with Connection Manager, 107
Copy Formatted command, independent bibliography, 161*f*
Copy-paste method, independent bibliography, 161
Creating bibliographies. *See* Bibliographies
Cross references, creating
bibliographies, 171–173
Custom Installation, selecting, 24*f*
Customizing. *See specific topic*
CWYW. *See* Cite While You Write (CWYW)

Damaged library, recovering, 49–51
dialog box, 50*f*
Date
Internet, citing references from sources on, 182
manually entering reference data, 65
Default reference type, setting, 62–63
Define Term Lists, selecting, 67*f*
Deleting duplicate records, 91
Deleting references, 82–83
PDA computers, 204*f*
Dialog box, 50*f*
Direct export, 116–117
databases allowing direct export, 116*f*
described, 103
Ovid, 126–127
Selecting Tools, 117*f*
Downloading files, filers, and styles for EndNote, 229
Downloading references from Websites of Journals, 74–77
conversion dialog box, 77*f*
customizing, 77*f*
online article, 75*f*, 76*f*

Downloading references from Websites (*cont.*)
opening files, 76*f*
selecting article for download, 74*f*
Drag-and-drop method
independent bibliography, 161
Duplicate references, 88–91
checking for, 88, 89*f*
customizing settings for Find Duplicates command, 90
deleting duplicate records, 91
Find Duplicates command, 89*f*, 90

Edited electronic source box, 191*f*
Editing a reference
group editing of references. *See* Group editing of references.
PDA computers, EndNote for, 203, 204*f*
Edition
Internet, citing references from sources on, 181
E-mail messages, examples, 184–185
EndNote libraries, 35, 37–56
bibliographies, inserting references from EndNote Library into, 140–144
changing existing citations, 142–144
copying citations, 143
CWYW menu, 141*f*
deleting citations, 144
editing citations, 142
Find Citations Dialog, 141*f*
moving citations, 143
setting EndNote formatting preferences, 143*f*
unformatting citations, 142–143
Character Map program, 73*f*
chart type references, 68–72
copying between libraries, 45
creating new library, 41, 42*f*
creating new reference, 61–63
customizing reference types, 58–61
damaged library, recovering, 49–51
dialog box, 50*f*
default library, setting, 47–48
default reference type, setting, 62–63
defined, 37
displaying record number in library, 43
downloading references from Websites of Journals, 74–77
conversion dialog box, 77*f*
customizing, 77*f*
online article, 75*f*, 76*f*
opening files, 76*f*
selecting article for download, 74*f*

EndNote libraries (*cont.*)
- EndNote Library window, 40–41, 42*f*
- features of, 37–38
- fields to be displayed in library, setting, 48–49
- figure type reference, 68–72
- fonts, setting, 46–47
- importing references from other reference management programs, 78
- Insert Object command, 70*f*
- Insert Picture command, 69*f*
- manually entering reference data, 63–65
  - date, 65
  - formatting by EndNote, 63
  - guidelines, 64–65
  - journal names, 65
  - pages, 65
  - year, 65
- merging EndNote libraries, 51–52
- navigating library, 43–44
- opening a library, 39–40
- output style, 46*f*
- PDA computers
  - attaching a note to a reference, 205
  - Configure Handheld Sync., 198*f*
  - copying EndNote library, 197–200
  - default reference type, 210
  - deleting a reference, 204*f*
  - displaying library information, 209*f*
  - displaying references, 201*f*
  - editing a reference, 203, 204*f*
  - entering new reference, 202, 203*f*
  - "figure" type references, 199, 200*f*
  - limited number of references, copying, 199
  - opening reference, 202
  - scroll button behavior in Edit view, 210
  - searching references, 205, 206*f*
  - selecting EndNote library, 198*f*
  - selecting Info from library menu, 208*f*
  - selecting Preferences menu, 209*f*
  - selecting references, 200*f*
  - selecting reference type, 204*f*
  - sorting references, 202
  - splash screen, 210
  - statistics about PDA EndNote Library, viewing, 208–209
  - working with EndNote library on PDA, 201–206
- previewing reference, 44–45
- publishing EndNote library on the Web, 52–56
  - RefMan, 53, 54*f*
  - RefWorks, 52–53, 55–56*f*

EndNote libraries (*cont.*)
- reference fields, 58–63
  - adding, 61
  - deleting, 61
  - in different types, 59*f*
  - explained, 58
  - manually entering reference data, 64–65
- references, entering into, 57–78
  - chart type references, 68–72
  - creating new reference, 61–63
  - downloading references from Websites of Journals, 74–77
  - figure type reference, 68–72
  - importing references from other reference management programs, 78
  - manually entering reference data, 63–65
  - overview, 57–58
  - reference fields, 58–63
  - reference types, 58–63
  - special characters, entering into references, 72–74
  - spell-checking, 74
  - table type references, 68–72
  - term lists, 65–72
- references, managing. *See* References, managing.
- reference types, 58–63
  - adding field to, 61
  - adding new, 61
  - choosing, 62
  - customizing reference types, 58–61
  - deleting field from, 61
  - explained, 58
  - hiding, 61
  - modification of, 59
  - reference fields in, 59*f*
  - screen to edit, 60*f*
  - setting default, 62–63
  - various types available, 59*f*
- RefMan, 53, 54*f*
- RefWorks, 52–53, 55–56*f*
- saving references to, connection file method, 114
- selecting article for download, 74*f*
- setting library preferences, 45–49
  - default library, setting, 47–48
  - fields to be displayed in library, setting, 48–49
  - fonts, setting, 46–47
- sorting library, 42, 44*f*
- special characters, entering into references, 72–74
  - Character Map program, 73*f*

EndNote libraries (*cont.*)
spell-checking, 74
starting EndNote program, 39*f*
table type references, 68–72
term lists, 65–68
"auto-completion," 66
basic features of, 66
defined, 66
Define Term Lists, selecting, 67*f*
helpful hints, 68
journal term lists, 66–67
setting preferences, 68*f*
"suggest terms as you type," 66
turning off, 68, 69*f*
unchecking "Read-only" box, 41*f*
working with, 38–45
copying between libraries, 45
creating new library, 41, 42*f*
displaying record number in library, 43
EndNote Library window, 40–41, 42*f*
navigating library, 43–44
opening a library, 39–40
previewing reference, 44–45
sorting library, 42, 44*f*
starting EndNote program, 39*f*
unchecking "Read-only" box, 41*f*
EndNote Styles, 35
Entering new reference
EndNote libraries. *See* EndNote libraries.
PDA computers, EndNote for, 202, 203*f*
Export method, 84–85
independent bibliography, 162

Field codes, bibliographies, 152–154
publisher, sending paper to
hidden field codes, citation with, 154*f*
removing field codes, 157
showing field codes, citation with, 155*f*
Field shading, turning off, 172*f*, 173
Figure references in manuscripts, working with, 149–152, 153*f*
CWYW setting for Figures and Tables, 154*f*
PDA computers, 199, 200*f*
punctuation for figures and tables, 153*f*
Figure type reference, 68–72
File compatibility issues, 35
File Locations Startup folder, 27*f*
Find Citations Dialog, 141*f*
Find Duplicates command, 89*f*, 90
Footnotes. *See* Notes
Format Bibliography, 136*f*, 145*f*
Freeware/shareware reference management software programs, 12*t*
FTP sites, examples, 186

Generate Figure List command, 146
Getting started with EndNote, 15–35
automatically updating, 33–34
file compatibility issues, 35
hand-held computer requirements, 17–18
hardware requirements, 15–16
installing EndNote, 19–22
checking installation, 22–33
obtaining EndNote, 18–19
operating system (OS) requirements, 16–17
overview of working with EndNote, 15, 16*f*
pricing for EndNote, 18*t*
technical requirements, 15–18
hand-held computer requirements, 17–18
hardware requirements, 15–16
operating system (OS) requirements, 16–17
word processor compatibility, 17
trial version, 19
uninstalling EndNote, 35
word processor compatibility, 17
Graphics, online
citing references from Internet sources, 184
Group editing of references, 91–96
change fields, 94, 95*f*
Change Text command, 92–93
move fields, 94, 95*f*, 96
record number, editing EndNote preferences to display, 91*f*

Hand-held computers
reference management software program, 9
requirements for EndNote, 17–18
Hardware requirements for EndNote, 15–16
Harvard style of referencing, 3
Help, checking installation, 34*f*
"Hide file extensions for known file types," unchecking, 29*f*
Hiding references, 83–84
HotSync operation, 197
Hyphenation command
Internet, citing references from sources on, 181*f*

Import filters, 114–116
defined, 114
described, 102–103

Import filters (*cont.*)
opening Filter Manager, 115*f*
PubMed, 120–124
using, 116
window, 115*f*
Independent bibliography
Copy Formatted command, 161*f*
copy-paste method, 161
creating, 160–162
defined, 160
drag-and-drop method, 161
export method, 162
print method, 162
Insert Object command, 70*f*
Insert Picture command, 69*f*
Installing EndNote
About EndNote screen, 22*f*
Backup folder, 20
checking installation. *See* Checking installation.
first time, for, 19
network, installing into, 13, 21
selecting About EndNote menu, 22
selecting backup of files, 21*f*
Selecting Start Run, 20*f*
typing in the "Run" box, 20*f*
upgrading from earlier version, 19–20
Internet
databases. *See* Internet databases, using EndNote with.
managing references from Internet sources. *See* Internet, citing references from sources on.
sources, citing references from. *See* Internet, citing references from sources on.
Internet, citing references from sources on, 175–192
authors, guidelines, 178
blank electronic source box, 191*f*
challenges, 176–177
customizing EndNote to cite references, 188–192
dates, guidelines, 182
definitions of common terms, 191–192*t*
edited electronic source box, 191*f*
edition, guidelines, 181
e-mail messages, examples, 184–185
examples of references, 182–187
e-mail messages, 184–185
FTP sites, 186
listserv message, 185
newsgroup message, 186
Internet, citing references from sources on (*cont.*)
examples of references (*cont.*)
software programs, 187
Web discussion forum posting, 185
World Wide Web (WWW), sources on, 182–184
FTP sites, examples, 186
general principles, 177
guidelines, 177–182
authors, 178
dates, 182
edition, 181
Hyphenation command, 181*f*
page information, 182
title, 178–179
URL, 179–181
Hyphenation command, guidelines, 181*f*
in-text citations, 182
listserv message, example, 185
location of information not static, 176–177
managing references from Internet sources, 187–192
blank electronic source box, 191*f*
customizing EndNote to cite references, 188–192
definitions of common terms, 191–192*t*
edited electronic source box, 191*f*
spelling of common terms, 191–192*t*
style manager, opening, 189*f*
Vancouver style of referencing, 189, 190*f*
missing citation elements, 176
newsgroup message, example, 186
page information, guidelines, 182
publication date unavailable, 176
software programs, examples, 187
spelling of common terms, 191–192*t*
title, guidelines, 178–179
URL, guidelines, 179–181
Vancouver style of referencing, 189, 190*f*
Web discussion forum posting, examples, 185
World Wide Web (WWW), examples, 182–184
online abstract, 183
online book, 183
online book chapters, 184
online databases, 184
online graphics, 184
online journal article, 183
online magazine article, 183
online newspaper article, 183
Web-Sites, 184

Internet databases, using EndNote with, 101–129
biomedical databases, commonly used, 102*t*
connection file method, 102, 106–114
choice of, 104–105
defined, 106
downloading connection files from Internet, 107–109
empty retrieved references window, 111*f*
establishing connection to database, 110
"favorite" connection files, setting, 107
illustrated, 108*f*
number of references found by search, window displaying, 113*f*
opening Connection Manager, 106*f*
PubMed, 110*f*, 118–119
replacing old connection file warning, 109*f*
retrieving references, 111–114
saving connection file to computer, 109*f*
saving references to EndNote Library, 114
searching database, 111
search window with search terms, 112*f*
selecting connection file for download, 108*f*
using method, 109–114
working with Connection Manager, 107
deleting temporary Internet files in Internet Explorer, 118*f*
direct export, 103, 116–117
databases allowing direct export, 116*f*
Ovid, 126–127
Selecting Tools, 117*f*
import filters, 102–103, 114–116
defined, 114
opening Filter Manager, 115*f*
PubMed, 120–124
using, 116
window, 115*f*
methods of using, overview, 102–105
overview, 104*f*
Ovid, 126–129
direct export method, 126–129
Web of Science, 127–129
PubMed, 103*t*, 110*f*, 117–125
click Save to download, 123*f*
connection file method, use for, 118–119
defined, 117
icon on desktop, 124*f*
import filter method, 120–124, 125*f*
performing search, 121*f*
saving NLM filter, 121*f*
selecting File from Send to menu, 122*f*
Internet databases, using EndNote with (*cont.*)
PubMed (*cont.*)
selecting MEDLINE, 122*f*
selecting NLM filter to download, 120*f*
selecting references, 121*f*
selecting type and location, 123*f*
Show All references, 125*f*
Internet Explorer
deleting temporary files, 118*f*

Journal names, manually entering reference data, 65
Journal term lists, 66–67

Launching EndNote Search, 85
Linking references
files, to, 96–98
OpenURL Link command, 98–99
Websites, to, 98
Listserv message, example, 185

Macro Security dialog box, 34*f*
Macro Security Warning, 31, 32*f*
Managing references. *See* References, managing
Manually entering reference data. *See* EndNote libraries
Manuscript, bibliographies
creating, 139–40
inserting references from EndNote library into, 140–144
changing existing citations, 142–144
copying citations, 143
CWYW menu, 141*f*
deleting citations, 144
editing citations, 142
Find Citations Dialog, 141*f*
moving citations, 143
setting EndNote formatting preferences, 143*f*
unformatting citations, 142–143
Manuscript templates, 139–140*f*
MEDLINE, 117
selecting, 122*f*
Microsoft Word, 18
checking installation, 23, 24*f*, 25
Macro Security dialog box, 34*f*
Macro Security Warning, 31, 32*f*
Security Level tab, Medium setting, 32*f*
shortcuts not working, 31
troubleshooting, 25–31
Macro Security Warning, 31, 32*f*
Move fields, 94, 95*f*, 96

National Library of Medicine, 117, 119
Network, installation, 13, 21
New England Journal of Medicine, 5
Newsgroup message, example, 186
Notes
  citing references in, 157–159
    customizing citations, 159–160
  inserting in list of references, 148–149

Online resources
  learning more about EndNote, 225
  writing for publication, 227–228
Opening references
  PDA computers, EndNote for, 202
OpenURL Link command, 98–99
Operating system (OS)
  requirements for EndNote, 16–17
Outline View, selecting, 147*f*
Output styles, bibliographies, 132–138
  authors' guidelines for biomedical journals, 133*t*
  Cite While You Write (CWYW), 132–133
  style manager, 134–138
    APA 5th style, 137, 138*f*
    Edit Citations dialog, 138*f*
    editing styles, 135–138
    examples of editing styles, 137–138
    favorites, marking styles as, 135
    Format Bibliography command, 136*f*
    overview, 134
    Style Editor window, 136*f*
    window, 135
    working in, 134
Ovid, 126–129
  direct export method, 126–129
  Web of Science, 127–129

Page information
  Internet, citing references from sources on, 182
Page number, inserting
  bibliographies, 170
Pages, manually entering reference data, 65
Palm, Inc., 18, 195
  application, installing EndNote for, 196–197
PDA computers, EndNote for, 193–210
  beaming references between PDAs, 207–208
  EndNote library
    attaching a note to a reference, 205
    Configure Handheld Sync., 198*f*
    copying, 197–200
    default reference type, 210

PDA computers, EndNote for (*cont.*)
  EndNote library (*cont.*)
    deleting a reference, 204*f*
    displaying library information, 209*f*
    displaying references, 201*f*
    editing a reference, 203, 204*f*
    entering new reference, 202, 203*f*
    "figure" type references, 199, 200*f*
    limited number of references, copying, 199
    opening reference, 202
    scroll button behavior in Edit view, 210
    searching references, 205, 206*f*
    selecting EndNote library, 198*f*
    selecting Info from library menu, 208*f*
    selecting Preferences menu, 209*f*
    selecting references, 200*f*
    selecting reference type, 204*f*
    sorting references, 202
    splash screen, 210
    statistics about PDA EndNote Library, viewing, 208–209
    working with EndNote library on PDA, 201–206
  HotSync operation, 197
  icon, EndNote, 197*f*
  installing
    HotSync operation, 197
    Palm application, 196–197
    and using, steps involved, 195*f*
  Palm application, 196–197
  statistics about PDA EndNote Library, viewing, 208–209
  technical requirements, 194–195
PDF (Portable Document Format), 18
Performing search, 88
Personal Digital Assistants. *See* PDA computers, EndNote for
Portable Document Format (PDF), 18
Pricing for EndNote, 18*t*
Print method
  independent bibliography, 162
Publisher, sending paper to, 152–157
  field codes, 152–154
    hidden, citation with, 154*f*
    showing, citation with, 155*f*
  generally, 157
  removing field codes, 157
  Toggle Field Codes, 155*f*
  traveling library, 155–156
Publishing EndNote library on the Web, 52–56
  RefMan, 53, 54*f*
  RefWorks, 52–53, 55–56*f*

PubMed, 103*t*, 110*f*, 117–125
click Save to download, 123*f*
connection file method, use for, 118–119
defined, 117
icon on desktop, 124*f*
import filter method, 120–124, 125*f*
performing search, 121*f*
saving NLM filter, 121*f*
selecting File from Send to menu, 122*f*
selecting MEDLINE, 122*f*
selecting NLM filter to download, 120*f*
selecting references, 121*f*
selecting type and location, 123*f*
Show All references, 125*f*
Punctuation for figures and tables, 153*f*

Record number, editing EndNote preferences to display, 91*f*
Recovery. *See* Damaged library, recovering
Reference, bibliography, and citation, 1–5
basic concepts, 1–2
common referencing styles, 3–4
effective reference management, information technology of, 5
Harvard style of referencing, 3
reference list and bibliography, 2–3
Vancouver style of referencing, 3–4
Reference fields. *See* EndNote libraries
Reference management software program, 7–13
bibliographies, creating, 9
choosing, 13
defined, 7
ease of use vs. functionality, 13
freeware/shareware reference management software programs, 12*t*
functions of, 8–9
handheld computer, working with, 9
individual use, 13
network installation, 13
platform, 13
price, 13
searching and retrieving references from online databases, 9
storing and managing references, 8
user survey, 8*f*
various programs, described, 10–11*t*
References, entering into EndNote libraries. *See* EndNote libraries
References, managing, 79–99
deleting references, 82–83
duplicate references, 88–91
checking for, 88, 89*f*
References, managing (*cont.*)
duplicate references, (*cont.*)
customizing settings for Find Duplicates command, 90
deleting duplicate records, 91
Find Duplicates command, 89*f*, 90
exporting references, 84–85
group editing of references, 91–96
change fields, 94, 95*f*
Change Text command, 92–93
move fields, 94, 95*f*, 96
record number, editing EndNote preferences to display, 91*f*
hiding references, 83–84
linking references
files, to, 96–98
OpenURL Link command, 98–99
Websites, to, 98
opening references, 82
reference window, understanding, 79–80
reverting references, 83
saving references, 82
searching references, 85–88
Boolean operators, 87*t*
comparison operators, 86*t*
launching EndNote Search, 85
performing search, 88
search window, 85–86
selecting references, 80–82
multiple references, 80–81
Select All, 81*f*
single references, 80
showing references, 83–84
working with references, 80–83
deleting references, 82–83
opening references, 82
reverting references, 83
saving references, 82
selecting references, 80–82
Reference types. *See* EndNote libraries
Reference window, understanding, 79–80
RefMan, 53, 54*f*
RefViz, use of with EndNote, 211–224
activating RefViz, 213–214
advisor, 223
column fields, selecting, 221*f*
Data Visualization command, 216*f*
definition of RefViz, 211–212
downloading RIS, 217*f*
exporting references, 224
galaxy view, 217, 218*f*
getting RefViz, 213
Inactive Data Visualization Command, 214*f*

RefViz, use of with EndNote (*cont.*)
installing
checking installation, 214*f*
technical requirements, 213
keyword tool, 222, 223*f*
location, screen to verify, 215*f*
matrix view, 217, 218
processes, flowchart, 212*f*
reference viewer, 219, 220–221*f*
search tool, 222
selecting RefViz.exe file, 215*f*
starting RefViz, 216
working in RefViz, 216–224
RefWorks, 52–53, 55–56*f*
Reverting references, 83

Saving references, 82
Scroll button behavior in Edit view, 210
Searching references, 85–88
Boolean operators, 87*t*
comparison operators, 86*t*
launching EndNote Search, 85
online databases, 9
PDA computers, 205, 206*f*
performing search, 88
search window, 85–86
Search window, 85–86
Security Level tab, Medium setting, 32*f*
Selecting article for download, 74*f*
Selecting references, 80–82
multiple references, 80–81
Select All, 81*f*
single references, 80
Selecting reference type
PDA computers, 204*f*
Sharing document with others, 155–156
"Show hidden files and folders," selecting, 30*f*
Showing references, 83–84
Software
programs, examples, 187
reference management. *See* Reference management software program.
Sorting references
PDA computers, EndNote for, 202
Special characters, entering into references, 72–74
Spell-checking, 74
Splash screen, PDA computers, 210
Startup folder, 27*f*, 28*f*
Style manager
opening

Style manager (*cont.*)
opening (*cont.*)
Internet, citing references from sources on, 189*f*
output styles, bibliographies, 134–138
APA 5th style, 137, 138*f*
Edit Citations dialog, 138*f*
editing styles, 135–138
examples of editing styles, 137–138
favorites, marking styles as, 135
Format Bibliography command, 136*f*
overview, 134
Style Editor window, 136*f*
window, 135
working in, 134
Subject bibliography, creating, 163–168
customizing, 165, 166*f*
definition of subject bibliography, 163
generally, 163–165
printing, 165–166
sample subject bibliography, 165*f*
saving, 165–166
selecting subject fields, 164*f*
selecting subject terms, 164*f*
Subject list, creating in bibliographies, 163–168
definition of subject list, 163
generally, 166–168
"Suggest terms as you type," 66

Table/chart references in manuscripts, working with, 149–152
CWYW setting for Figures and Tables, 154*f*
punctuation for figures and tables, 153*f*
working with table type references, 68–72
Technical requirements for EndNote, 15–18
hand-held computer requirements, 17–18
hardware requirements, 15–16
operating system (OS) requirements, 16–17
word processor compatibility, 17
Term lists. *See* EndNote libraries
Title
Internet, citing references from sources on, 178–179
Toggle Field Codes, 155*f*
Traveling library, 155–156
Trial version of EndNote, 19
Troubleshooting, checking installation, 25–31
Trust All . . . box, 33*f*

Uninstalling EndNote, 35
Updating, automatically, 33–34

URL
Internet, citing references from sources on, 179–181

Vancouver style of referencing, 3–4, 189, 190*f*

Web discussion forum posting, examples, 185
Web of Science, 127–129
Websites, linking references to, 98
Windows Explorer
folder options, selecting, 29*f*
launching, 28*f*
Word. *See* Microsoft Word
Word processors
checking installation, word processor support, 23–33
Word processors (*cont.*)
compatibility, 17
World Wide Web (WWW)
citing references from, 182–184
online abstract, 183
online book, 183
online book chapters, 184
online databases, 184
online graphics, 184
online journal article, 183
online magazine article, 183
online newspaper article, 183
Web-Sites, 184

Year
manually entering reference data, 65

Printed in the United States of America